INTR

Genetics

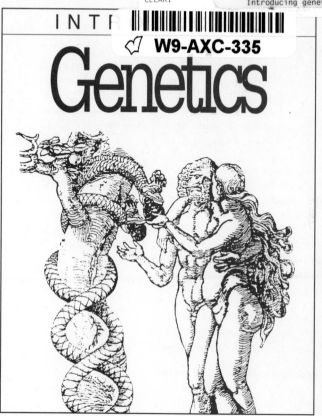

Steve Jones and Borin Van Loon

Edited by Richard Appignanesi

ICON BOOKS UK   TOTEM BOOKS USA

This edition published in the UK
in 2000 by Icon Books Ltd.,
Grange Road, Duxford,
Cambridge CB2 4QF
email: icon@mistral.co.uk
www.iconbooks.co.uk

Distributed in the·UK, Europe,
Canada, South Africa and Asia by the
Penguin Group: Penguin Books Ltd.,
27 Wrights Lane, London W8 5TZ

This edition published in Australia
in 2000 by Allen & Unwin Pty. Ltd.,
PO Box 8500, 9 Atchison Street,
St. Leonards NSW 2065

Previously published in the UK and
Australia in 1993 under the title
Genetics for Beginners

Reprinted 1994, 1996, 1998

First published in the United States
in 1994 by Totem Books
Inquiries to: PO Box 223,
Canal Street Station,
New York, NY 10013

Reprinted 1998

In the United States,
distributed to the trade by
National Book Network Inc.,
4720 Boston Way, Lanham,
Maryland 20706

Originating editor: Richard Appignanesi

Printed and bound in Australia
by McPherson's Printing Group, Victoria

Genetics is about differences…

It is also about similarities — between relatives, alive or dead

ELIZABETH II

ALBERT

KAISER BILL

VICTORIA

4

and between different creatures,
living or extinct.

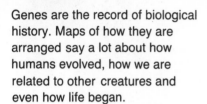

Genes are the record of biological
history. Maps of how they are
arranged say a lot about how
humans evolved, how we are
related to other creatures and
even how life began.

Much of genetics is
geography, on one scale
or another.

But genetics began long after the world was explored, . . .

. . . and later than any other biological science — because, unfortunately, the obvious has usually turned out to be wrong.

For a thousand years people believed that relatives look alike because they shared the same environment and that experience changes the way you look.

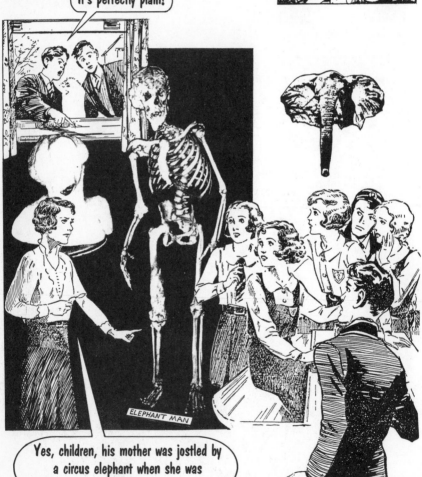

YOU'RE SO OBVIOUS THAT YOU **MUST BE** WRONG, DAHLING!

It's perfectly plain!

Yes, children, his mother was jostled by a circus elephant when she was expecting.

It must be true — it's in the Bible.

But children don't inherit what their parents experienced.

Well, if that doesn't work, perhaps children are the **average** of what went before. Darwin liked the idea that children were formed by mixing the blood of their parents. After all, his own family was pretty blue-blooded.

Soon he read a nasty little article by Fleeming Jenkin, a Scots engineer. It pointed out a fatal flaw: if inheritance works like this, then any favourable character will be diluted out each generation until it disappears. The theory of evolution would not work! Jenkin had typically racist views. . .

Soon, Darwin's cousin, Francis Galton, got interested. Galton was a strange, unlikeable man.

My cousin's a genius, and so am I!

Like most Victorian scientists, he was rich. Unlike his cousin he completed his medical course (although he never practised). During it, he tried every drug in the book in alphabetical order, giving up at the purgative Croton Oil.

HEREDITARY GENIUS

11

Galton travelled in Africa, riding on a bull into a chief's house to frighten him into submission and measuring his wives' buttocks with his naval sextant. He was interested in the inheritance of "genius" (judges were one example).

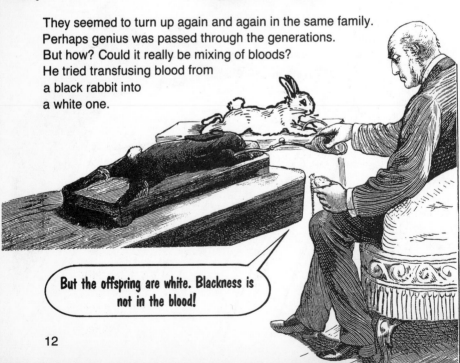

They seemed to turn up again and again in the same family. Perhaps genius was passed through the generations. But how? Could it really be mixing of bloods? He tried transfusing blood from a black rabbit into a white one.

But the offspring are white. Blackness is not in the blood!

Galton died, childless, in 1911. He left a fortune to found the Laboratory for National Eugenics at University College London.

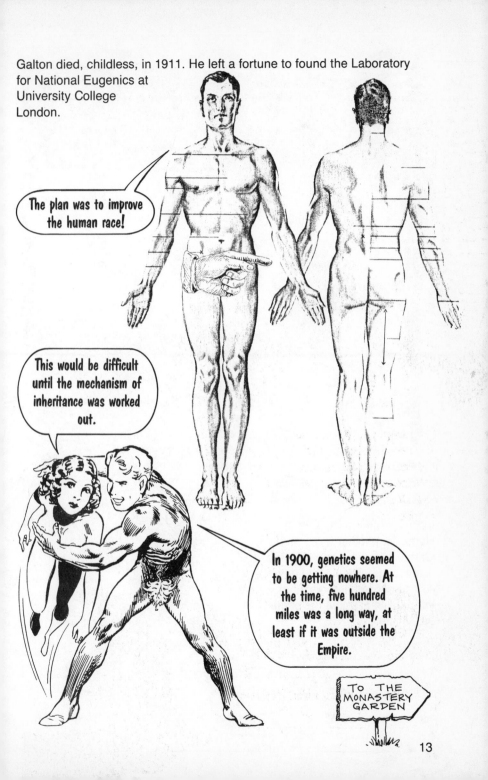

The plan was to improve the human race!

This would be difficult until the mechanism of inheritance was worked out.

In 1900, genetics seemed to be getting nowhere. At the time, five hundred miles was a long way, at least if it was outside the Empire.

TO THE MONASTERY GARDEN

THE AUGUSTINIAN MONASTERY OF ST. THOMAS

In Brünn, now in the Czech Republic, another failed student — Gregor Mendel, who had studied science at university but gave up — had become interested in inheritance at about the same time as Galton.

He had more sense than Galton; he studied not humans but peas. They had all kinds of advantages — clean, easy to keep, and the divorce rate was low. What's more, each plant was both male and female, and could fertilize itself.

Well, if I can't have sex, at least the peas can.

So THIS is what they mean by sex with someone you really love.

15

Farmers had bred many different **pure lines** of peas: within a line every plant was the same; between lines they were different.

Mendel realized that this is just what was needed to study how inheritance works. He fertilized seeds from a line with round peas with pollen from another line with wrinkled seeds.

All the offspring of this cross were round — not the average of their parents at all, but like only one of them.

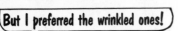
But I preferred the wrinkled ones!

Then, Mendel grew up these round peas, and self-fertilized them; he put pollen onto the eggs of the same plant. When he grew these up — a big surprise: both types, round and wrinkled, came back again!

What's more, there were always three round peas to one wrinkled.

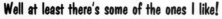

Well at least there's some of the ones I like!

Mendel had . . .

THE BIGGEST IDEA IN THE HISTORY OF GENETICS

Perhaps there's more to peas than the way they look! They might carry some concealed instructions which do not always reveal what they say; a round pea could contain hidden within itself the instruction for wrinkles.

Mendel suggested that pollen and egg each carried a particle (called a gene nowadays) containing the code for the shape of the pea in their offspring.

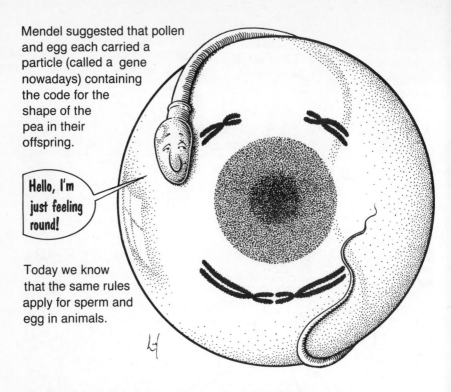

Hello, I'm just feeling round!

Today we know that the same rules apply for sperm and egg in animals.

When pollen met egg, the offspring had two particles or genes. Sometimes, one particle hid the effects of the other.

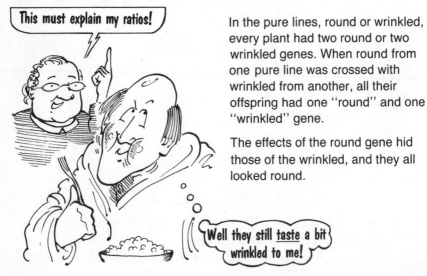

This must explain my ratios!

In the pure lines, round or wrinkled, every plant had two round or two wrinkled genes. When round from one pure line was crossed with wrinkled from another, all their offspring had one "round" and one "wrinkled" gene.

The effects of the round gene hid those of the wrinkled, and they all looked round.

Well they still taste a bit wrinkled to me!

Round was **dominant** to wrinkled, which was **recessive**.

In the next generation all these round peas had two different genes. They therefore made two kinds of pollen or eggs; half with round and half with wrinkled particles or genes.

When they were self-fertilized, one time in four, "round" pollen met "round" egg; another time in four "wrinkled" met "wrinkled"; and twice in every four, "round" and "wrinkled" got together to give a round pea. Add them together, and Mendel's magic three to one ratio of round to wrinkled peas in the next generation was explained!

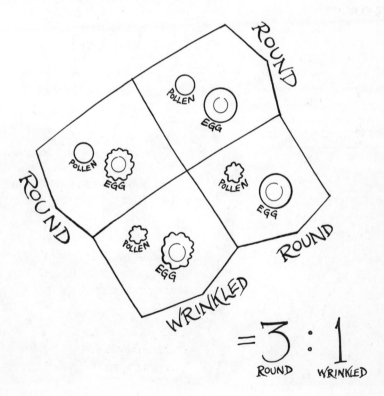

$= 3 : 1$
ROUND   WRINKLED

Mendel did it again with yellow and green peas, or with tall and short. He got the same result each time; and it worked for every character he chose. What's more, the shape of the pea made no difference to how colour was inherited. The genes were independent.

Genetics was, it seemed, based on particles passed from parents to offspring. It all seemed so simple.

Alas, it wasn't. He moved on to study other plants which have complicated patterns of inheritance; and his laws seemed to fall to pieces. Like Galton and Darwin, he suffered from attacks of depression, and withdrew into administration.

Mendel's paper *Versuche über Pflanzenhybriden* (*Experiments on Plant Hybridization*) was published in 1866 in a little-known journal, the *Transactions of the Brünn Natural History Society*.
He sent it to the most eminent biologists
of his day, but it was ignored.

This Mendelism is a damp squib!

NAGELI, PROFESSOR OF BOTANY

VERSUCHE ÜBER PFLANZENHYBRIDEN
MENDEL

They were interested in a much bigger question. Now we know that they were asking the right question at the wrong time. They had no chance of answering it — and it remains unsolved. How does a fertilized egg with no structure of its own develop into the incredible complexity of a human being — or a pea?

UNINTERESTING TOSH!

The mechanism of inheritance seemed much less interesting to the aristocrats of Victorian biology. Mendel had found the right answer to a simpler question — and was ignored for his pains. In 1900 his work was rediscovered. It explained inheritance in all kinds of plants and even in mice and chickens.

Soon everyone was doing it. Humans came next.

Obviously, humans cannot be crossed together like peas. Or can they? Frederick the Great of Prussia had been quite successful in mating tall men with tall women to get some impressive guards for his palace.

Most of the time, human genetics has to wait for nature's experiments. People choose their own mates.

I've always fancied big women.

23

Often, family history is recorded in a pedigree — from the French **pied de grue**, crane's foot, after the way the lines splayed out from the centre in some ancient family trees.

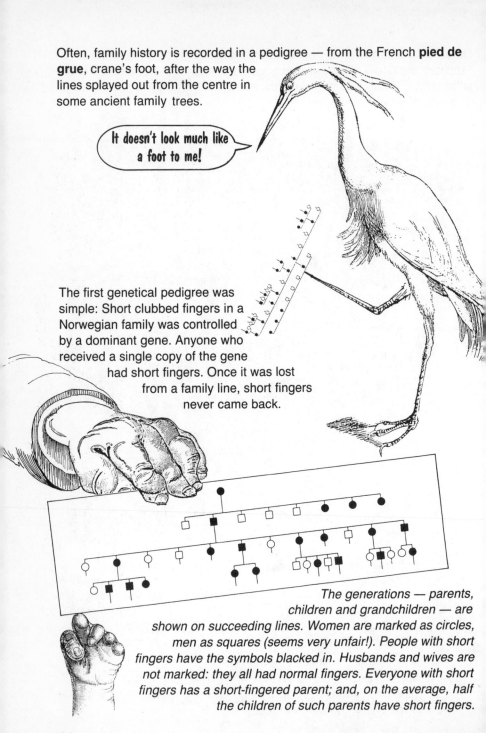

It doesn't look much like a foot to me!

The first genetical pedigree was simple: Short clubbed fingers in a Norwegian family was controlled by a dominant gene. Anyone who received a single copy of the gene had short fingers. Once it was lost from a family line, short fingers never came back.

The generations — parents, children and grandchildren — are shown on succeeding lines. Women are marked as circles, men as squares (seems very unfair!). People with short fingers have the symbols blacked in. Husbands and wives are not marked: they all had normal fingers. Everyone with short fingers has a short-fingered parent; and, on the average, half the children of such parents have short fingers.

Dominant characters seemed simple enough. Soon, characters controlled by recessive genes began to turn up. Only those inheriting two copies of the gene, one from each parent, showed its effects.

Such traits cropped up in families, often skipping generations. This explained an old problem — atavism; the tendency of a child to resemble a remote relative or distant ancestor.

The first case of recessive inheritance was albinism: Most parents of an albino child were normal and if an albino married a normal person, their children usually had normal skins. The earliest known albino was Noah. As the Book of Enoch says, "his hair was white and fine as snow".

But Ham, Shem and little Japhet look just like you.

Just as in the peas, human genetics all looked terribly simple. Soon, all kinds of pedigrees appeared. Some were sensible and stood the test of time. Others were deeply foolish. Genes were (allegedly) discovered for love of the sea — thalassophilia — and even for outbursts of bad temper.

I can't help it, it's in my genes.

Human genetics was beset by cranks. They were convinced that it was their duty to rid the world of non-existent genes for idiocy or criminality.

In the 1920s, Galton's Laboratory for National Eugenics split: one branch became the Galton Laboratory at University College London and spent its time doing biological research. The other called itself the Eugenics Society, and for many years pursued a mission to improve the human race.

All kinds of unexpected people were involved. The great pioneer of family planning Marie Stopes — a member of the Eugenics Society — was driven by a desire to prevent the lower orders from having children and reducing the quality of the British nation.

WORKERS OF ALL LANDS DON'T UNITE

Other descendants of Galton had even less attractive views. Hitler was the most notorious, but there were plenty of others.

If we desire a certain type of civilization we must exterminate the sort of people who do not fit into it.

The unnatural and increasingly rapid growth of the feeble-minded and insane classes constitutes a national and race danger which is impossible to exaggerate. I feel that the source from which the stream of madness is fed should be cut off and sealed off before another year is past.

The quality of our hereditary endowment is a hundred times more important than the dispute between capitalism and socialism.

G.B. Shaw

W.S. Churchill

A. Hitler

While this went on, real genetics made progress behind the headlines. All kinds of questions began to turn up. Where does genetic variation come from in the first place? We are so used to it that we find similarity disturbing.

Why do some people have short fingers — and why, for that matter, are some peas round and others wrinkled? **Something** must have made them change. If inheritance was perfect, every living creature would be the same. Genetics would not exist and neither would evolution.

In 1901, the Dutchman de Vries was testing Mendel's laws in evening primroses. To his surprise, every now and again a new flower colour turned up even within a pure line — and was then inherited. He called these random changes in genes

...**mutations**

Perhaps looking at mutations — damage in the inherited machinery — would help in understanding how genes actually work.

An English doctor, Archibald Garrod, was studying a rare inherited disease, alkaptonuria.

The symptoms were alarming (but not very dangerous). After eating certain foods, the urine turned black and smelly.

In 1909 he found that the smelly substance was a chemical which appeared because of a failure to complete the breakdown of a certain item in the food.

It had long been known that the workings of the body depended on **enzymes**; chemical catalysts which speeded up metabolism. All enzymes were proteins.

Perhaps the symptoms of alkaptonuria arose because an enzyme was not working properly. Garrod suggested that genes made enzymes. Perhaps genes **were** enzymes — although he had no real evidence for this.

If that was — roughly — what Mendel's particles did, **where** were they? They must be passed on in sperm and egg; and there were lots of theories as to just where they might be.

At around the same time as Garrod, Thomas Hunt Morgan became interested in genetics. He was a professor at Columbia University in New York. When looking for something to study, he had a stroke of genius or, almost the same thing, good luck.

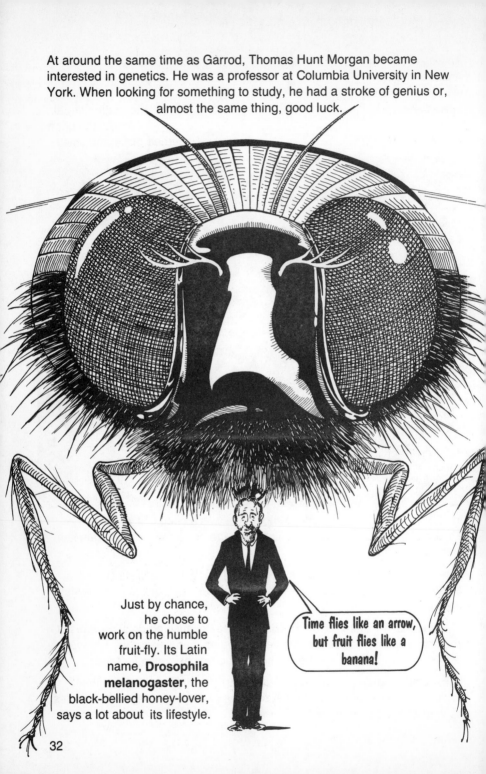

Just by chance, he chose to work on the humble fruit-fly. Its Latin name, **Drosophila melanogaster**, the black-bellied honey-lover, says a lot about its lifestyle.

Time flies like an arrow, but fruit flies like a banana!

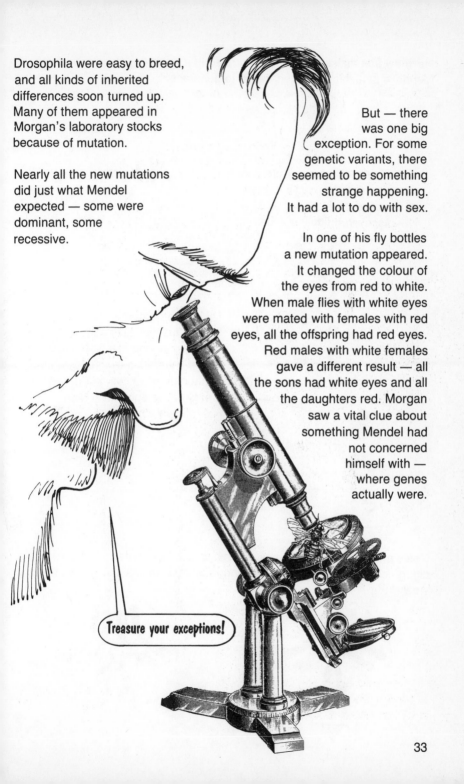

Drosophila were easy to breed, and all kinds of inherited differences soon turned up. Many of them appeared in Morgan's laboratory stocks because of mutation.

Nearly all the new mutations did just what Mendel expected — some were dominant, some recessive.

But — there was one big exception. For some genetic variants, there seemed to be something strange happening. It had a lot to do with sex.

In one of his fly bottles a new mutation appeared. It changed the colour of the eyes from red to white. When male flies with white eyes were mated with females with red eyes, all the offspring had red eyes. Red males with white females gave a different result — all the sons had white eyes and all the daughters red. Morgan saw a vital clue about something Mendel had not concerned himself with — where genes actually were.

Treasure your exceptions!

33

He knew that males and females differed in a way that was passed, like Mendel's particles, from generation to generation.

Every cell in every creature contains some string-like bodies, the chromosomes. They had been discovered fifty years before.

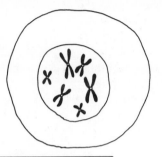

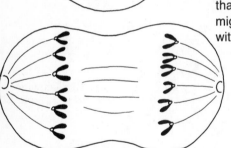

CHROMOSOMES DOUBLING & DIVIDING

Just like Mendel's hypothetical particles, they split and were divided among the offspring in the next generation — already a hint that chromosomes and genes might have something to do with each other!

Males and females were almost the same, but there was one important difference. Females had two "X" chromosomes, males a single X and a smaller Y.

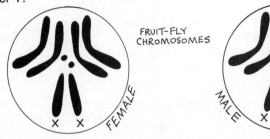

FRUIT-FLY CHROMOSOMES

FEMALE
X X

MALE
X Y

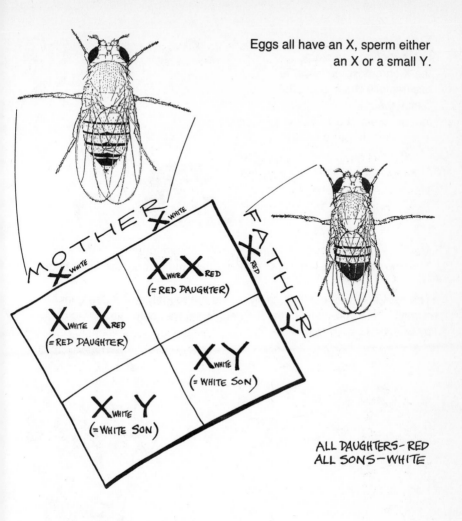

Eggs all have an X, sperm either an X or a small Y.

MOTHER

FATHER

X_{WHITE} X_{WHITE} X_{RED}

	X_{WHITE} X_{RED} (= RED DAUGHTER)
X_{WHITE} X_{RED} (= RED DAUGHTER)	X_{WHITE} Y (= WHITE SON)
X_{WHITE} Y (= WHITE SON)	

ALL DAUGHTERS - RED
ALL SONS - WHITE

Morgan noticed something important.
The pattern of inheritance of eye colour followed that of
the X. Sons got their X (and their eye colour) from their mother, and the Y
from their father; daughters got one X chromosome from each parent. The
Y chromosome did not seem to carry any genes for eye colour, so that any
gene on the X showed its effects. As red eye was dominant to white, red
father with white mother gave red daughters and white sons.

He suggested that the eye colour gene was **linked** to the X
chromosome. Presumably, this meant that the gene was actually on the
chromosome!

Soon there was a
final proof. In one stock,
the X chromosome became
accidentally stuck to another one.
Simultaneously, the pattern of
inheritance of the white eye gene changed.
The genes **must** be on the
chromosomes.

Mendel's particles had — approximately — been found.
The first faltering step towards making the genetic map had been taken.

The next step was obvious; to try to find out roughly where the genes
actually were in relation to each other. There was only one way to do it —
by breeding experiments.

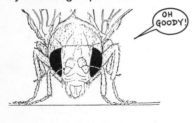

Back to T.H. Morgan and his fruitflies.

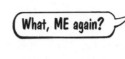

What, ME again?

Morgan and his followers soon found that plenty of genes — one for wing
length, for example — as well as that for white eye were on the X
chromosome. All these genes were **linked** to the X. This meant, of course,
that they were linked to each other. Within a linkage group, different genes
tended to be passed on together; in different groups they were
independent.

The number of linkage groups was the same as the number of chromosomes. It seemed that every chromosome had its own set of linked genes. For some genes, at least, Mendel was wrong: the inherited particles are not necessarily independent of each other.

But, soon, another complication. Linkage was not perfect. Even linked genes sometimes split from each other as the generations passed.

Is NOTHING perfect?

Morgan crossed flies with white eyes and short wings with other flies with red eyes and wings of normal length. To some extent, white and short went together in subsequent generations, as did red and normal. However, soon the two began to drift apart.

Darling, we seem to be drifting apart!

After many generations, there were plenty of flies with white eyes and normal wings, or with red eyes and short.

It was a bit like shuffling a new pack of cards again and again. Every time, a bit of the original order was lost:

just as when playing bridge, a good player can work out what the original arrangement must have been by comparing the order of cards each time the pack is shuffled.

In 1913 his student, A.H. Sturtevant, wrote a paper whose title summed up geneticists' views of the gene for the next seventy years — "The linear arrangement of six sex-linked factors in Drosophila, as shown by their mode of association". Sturtevant looked at lots of genes to see if they tended to travel together down the generations.

They do! And their fondness for each other varies greatly from gene to gene.

Soon he suggested that genes which nearly always went together were close together on the same chromosome; those less devoted to each other's company were further apart; and those which behaved independently were on different chromosomes. Mendel had been lucky — all the genes he looked at were on different chromosomes (or so far apart on the same one that he did not notice).

Slowly, Sturtevant and his successors built up a **linkage map** based on the strength of the tendency of different genes to travel down the generations together.

There was a clear pattern. As the map was built up from different small groups of genes, it became clear that within a single chromosome, all the genes were in order — they formed a chain of instructions arranged in a line.

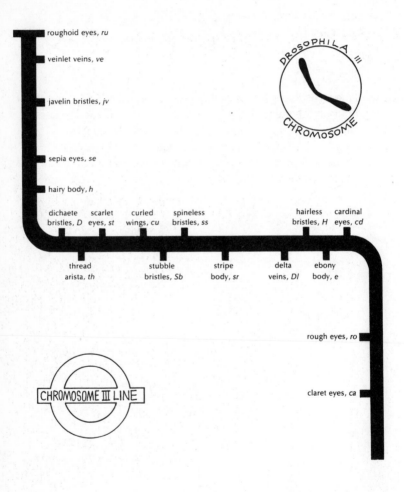

The same is true of every other creature, although the number of chromosomes varies enormously. The correct human chromosome number — 23 in sperm or egg, 46 in body cells — was not established until 1956.

Linkage mapping made great progress — quite soon, every known inherited variant in Drosophila had been mapped in this way. Of course, things were much worse for humans. Small families and unplanned matings made it almost impossible to make a genetic map.

Suddenly, politics raised its ugly head again.

Stalin didn't like the idea at all. He hated the idea that fate — even the fate of a fruit-fly's eye — was determined by biology.

Marx had insisted that

By changing the environment it is possible to do anything.

Mendelism and Morganism must be a capitalist plot!

His Director of Agriculture, Trofim Denisovitch Lysenko, started a hate campaign against genes and chromosomes. It went a long way; the whole of Soviet agriculture was planned on the theory that exposing parents to a new environment (a cold spring, for example) meant that the offspring would inherit an ability to cope with icy weather.

This was an expensive disaster for farming — and for genetics. Several geneticists were imprisoned. In 1940 Vavilov, one of the best, was arrested. Lysenko took his job as Director of the Institute of Genetics — where he stayed until 1962. After 1700 hours of interrogation Vavilov was found guilty in a five-minute trial of "belonging to a rightist conspiracy, spying for England, and sabotage of agriculture". In 1943 he died of starvation in a prison camp.

In spite of Stalin's fears, the genetic map was still pretty crude. Only the dim outline of the continents — the chromosomes — had been distinguished.

There was another big question in genetics, which the map itself said nothing about. Just where were these inherited particles? Fruit-flies gave the answer again. This time, the American geneticist T.H. Muller was involved. He remained an enthusiastic communist for much of his life.

He was interested in mutation. What made genes change from one form to another? Perhaps if he could find out he could get an idea of what they actually are.

Like Morgan, he used fruit-flies. He chose to study a simple kind of mutation — those which killed their carriers.

Lots of things seemed to increase the number of lethal mutations. Even a slight increase in temperature doubled the rate.

In 1930 he found that X-rays had a spectacular effect. A sudden burst to the parents could increase the mutation rate by a hundred and fifty times!

Soon, governments got very interested. Perhaps there were military implications! In the late 1930s, Charlotte Auerbach, a German refugee working in Edinburgh, set to work on chemicals. The war gases (mustard gas, for example) seemed a good place to start. They produced painful burns which took months to heal — just like a burst of radiation.

Sure enough, poison gases greatly increased the number of mutations — a result that was kept secret until the end of the war.

The gene began to look a bit like a target. Firing X-rays at it caused damage each time they struck — and the more X-rays, the more chance of hitting the target. The target must be a chemical substance of some kind, too.

But what kind of chemical?

Many years before, a German biochemist called Miescher had been interested in a strange substance found in the nuclei of cells — and very abundant in sperm and eggs.

To get hold of it, lots of cells were needed. Pus is a good place to look; it is full of white blood cells. Miescher visited septicaemia wards, and came back with bandages dripping with the precious substance.

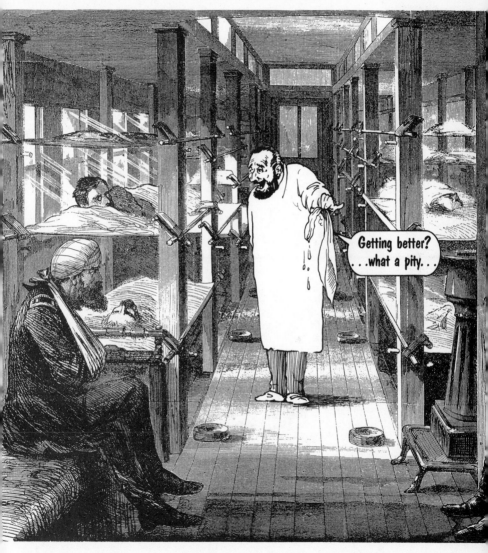

Miescher's material had something only found in the cell nucleus. He called this new substance "nucleic acid". The nucleus had lots of proteins, too. Perhaps one of the two had something to do with inheritance — and, if so, the proteins seemed the best bet.

Proteins are made of chemical building blocks called amino acids. There are twenty of these, and they differ greatly from each other in chemical structure. Proteins were intricate things — presumably, just like genes must be.

The nucleic acid seemed a much less promising candidate. It had only four kinds of building block, each chemically very similar. This scarcely seemed enough to contain the information passed on in the genes! For many years, nucleic acids were dismissed as the ''boring substance''.

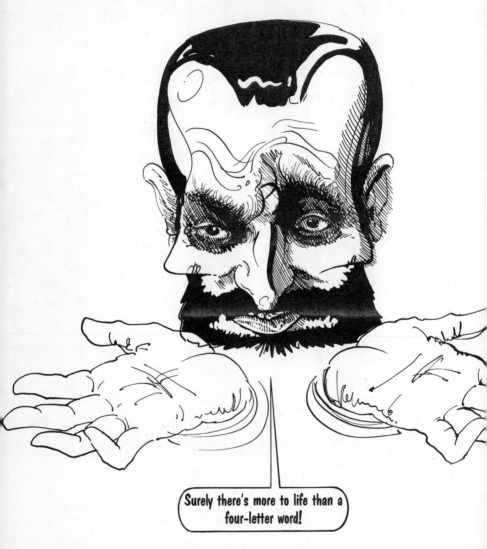

Surely there's more to life than a four-letter word!

But soon there was a strong hint that they really were involved in inheritance. In 1944 Avery, McLeod and McCarty were working on pneumonia — which still killed thousands of people (soldiers in particular).

When they grew the bacteria in the lab, they found that there were two kinds of colonies; just like the peas they were smooth or rough. Crosses between the two types could be made by injecting both forms into mice and looking at what emerged during the course of infection.

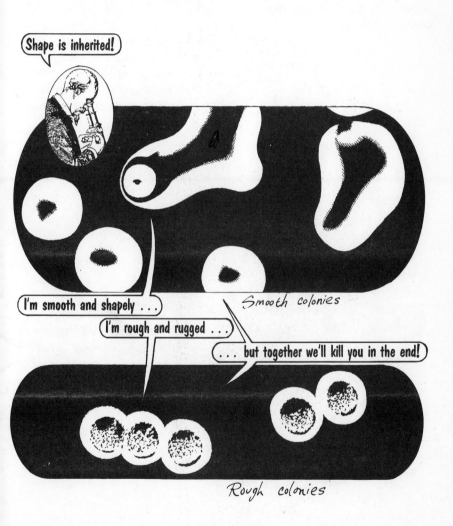

49

So bacteria had genes, too. There was soon

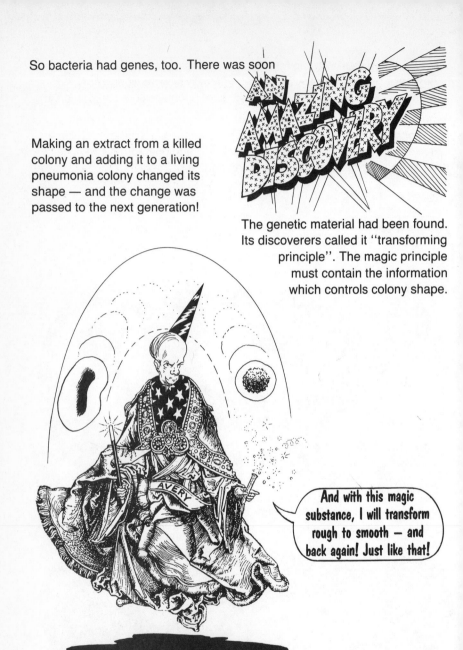

AN AMAZING DISCOVERY

Making an extract from a killed colony and adding it to a living pneumonia colony changed its shape — and the change was passed to the next generation!

The genetic material had been found. Its discoverers called it "transforming principle". The magic principle must contain the information which controls colony shape.

And with this magic substance, I will transform rough to smooth — and back again! Just like that!

AVERY

Transforming principle was, it soon turned out, a **nucleic acid** and not a protein.

Nucleic acids were everywhere. There were two main kinds, named according to the type of sugar with which they were associated.

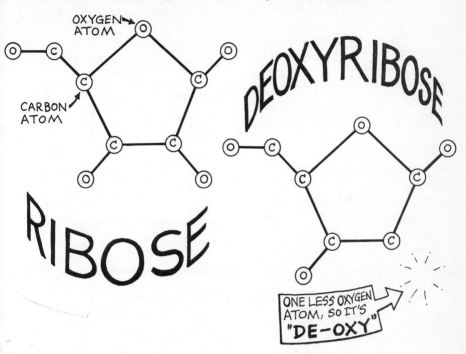

In creatures more advanced than bacteria — those which have a separate cell nucleus whose contents are passed on to the next generation — ribose nucleic acid (or RNA) is found in both the nucleus and the cytoplasm. The other kind, deoxyribose nucleic acid — DNA — is only in the nucleus.

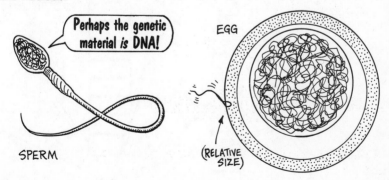

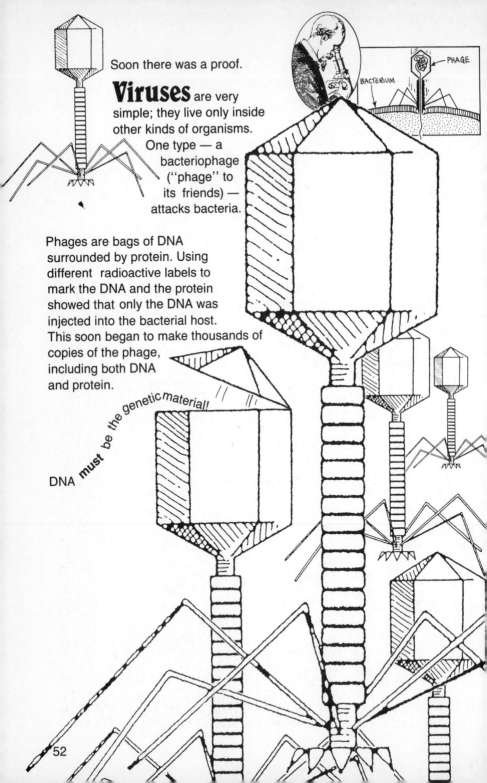

Soon there was a proof.

Viruses are very
simple; they live only inside
other kinds of organisms.
One type — a
bacteriophage
("phage" to
its friends) —
attacks bacteria.

PHAGE

BACTERIUM

Phages are bags of DNA
surrounded by protein. Using
different radioactive labels to
mark the DNA and the protein
showed that only the DNA was
injected into the bacterial host.
This soon began to make thousands of
copies of the phage,
including both DNA
and protein.

DNA **must** be the genetic material!

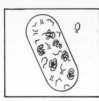 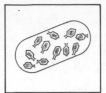 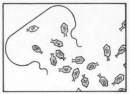

But how could such a simple substance copy itself and pass information from one generation to the next?

There was already a hint. There were only four units in DNA — Adenine, Guanine, Cytosine and Thymine (A, G, C and T for short). The number varied from creature to creature — but there was always the same proportion of A to T and G to C.

I say, Carruthers, this jolly old code seems to match the letters in pairs. Damned clever, what?

In the early 1950s an American biologist, James Watson, came to Cambridge. He was set to work on the biochemistry of nucleic acids, but did not like it. He soon met Francis Crick, a physics graduate from Galton's old institution, University College London.

Both of them were interested in the structure of biological molecules and hoped to use methods developed by physicists to study crystals. As they later admitted, they were dilettantes in the fiercely competitive world of crystallography.

They spent a *great* deal of time in a Cambridge pub, the Eagle.

ROSALIND FRANKLIN

Physics helped. If X-rays are shone at a crystal, some pass through and some bounce off. With some nifty mathematics the crystal's shape can be inferred. It is a bit like a blind snooker player shooting balls at random across a table. By measuring the angle at which the balls come back and counting those which never return, he can work out the shape and position of the pockets.

Rosalind Franklin, an intensely serious and skilled crystallographer, did a lot of the basics; but she was unlucky — she did not come up with the final structure. And, after all, she was only a woman! She died before the DNA story was complete.

Watson and Crick looked at the diffraction pattern which emerged when X-rays were beamed at the DNA. Several other people — including the famous American chemist Linus Pauling — were doing the same thing, often with more technical skill.

But they did not have a flash of insight at the right time.

One day in 1953, Watson and Crick saw that the best explanation of the patterns which emerged from the X-rays was a double helix — a structure a bit like a spiral staircase. Suddenly, everything began to fit.

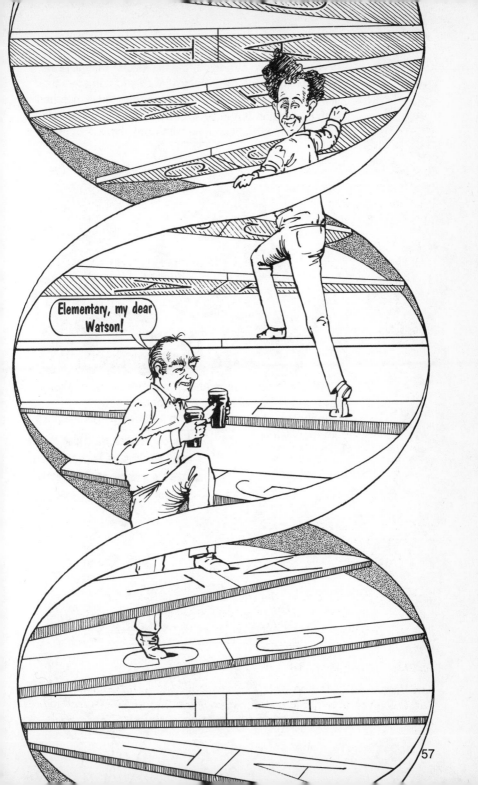

57

Perhaps the two strands of the helix were held together because different bases paired with each other — a bit like playing dominoes: one number had to be matched with its partner. In the case of DNA, Adenine was paired with Thymine, and Guanine with Cytosine.

My T matches your A.

TIME GENTLEMEN, PLEASE!

DNA's structure hinted at how it reproduced. As Watson and Crick said, with a certain smugness, when announcing the double helix in a paper in the scientific journal *Nature* in 1953:

It has not escaped our notice that the specific pairing we have postulated suggests a possible copying mechanism for the genetic material.

TWEEDLE WAT

TWEEDLE CRICK

Five years later, two American bacteriologists, Meselson and Stahl, showed that they were right. Bacterial DNA was labelled with a heavy chemical in the food.

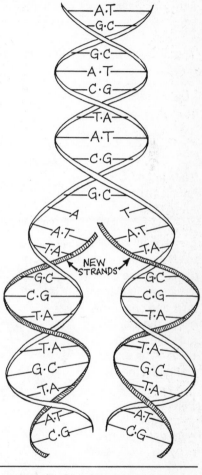

NEW STRANDS

Each generation from then on, the bugs were fed on ordinary, light food.

Their DNA was weighed by seeing how easily it floated. Each generation, an original heavy DNA strand remained, but was gradually outnumbered by more and more light strands. The primal strand persisted, making copies of itself; and each of its daughters did the same thing. One DNA chain was acting as a template to make another. This method of copying conserved just a part of the original structure.

Apart from some details — such as a series of specialized enzymes or **polymerases** which are now known to be involved — that was the copying sorted out. But that still leaves an important question unanswered: how is the genetic information coded into the DNA?

Everybody assumed that DNA was the same as the chromosome map on a much smaller scale — a line of letters arranged in order, containing the instructions to make a fruit-fly — or a human being.

The next job was to decipher the code.

Galton would have been proud of us — we're both hereditary geniuses!

Watson and Crick struck again. They knew that it was possible to mutate genes in all kinds of ways — to change them by damaging the DNA structure. Radiation, heat and chemicals all did the job. It was a bit like playing darts — an expert could aim the mutations at a particular gene and usually hit in the right general area.

In this game only the triples count

My God, Watson, when are you going to let me stop playing these stupid games?

Some chemicals did the damage in a strange way — they actually inserted themselves into the DNA message. Others knocked single A, G, C or T letters out of it.

Watson and Crick aimed their chemical darts at the DNA of some bacteriophages. When there was only a single hit, the phages could not grow. A double hit had the same effect. However, if **three** extra letters were inserted into the DNA message, the phages grew almost normally.

They suggested that the DNA code was read off in groups of three letters from one end to the other. If there were one or two extras, the message from then on was garbled. If there were three, it began to make sense again.

The genetic information was in a simple language, with three-letter words and an alphabet of four distinct letters.

All this coding went on in the cell's control centre — the nucleus. Most of the hard work of actually making the proteins happened in the rest of the cell. How was the information transferred from the management to the shop floor (or, in the Eagle, from the customers to the bar)?

Another nucleic acid, ribose nucleic acid, or RNA, was involved. It came in several flavours. One, the messenger RNA, took instructions from the DNA in the nucleus to the production line — which was made of ribosomal RNA. On the line, a series of specialized workers, the transfer RNAs, picked up components for making the protein and bolted them together.

The process of reading the message from DNA into mRNA is called **transcription**; actually making the protein is known as **translation**. Antiobiotics such as streptomycin and bacterial poisons such as diphtheria toxin work by blocking one or the other.

Never short of self-confidence, Watson called this the "Central Dogma" of molecular biology:

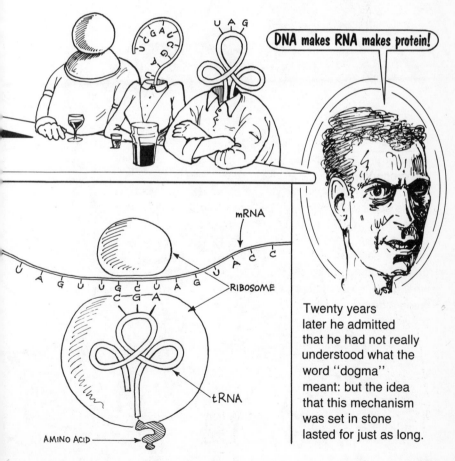

DNA makes RNA makes protein!

mRNA

RIBOSOME

tRNA

AMINO ACID

Twenty years later he admitted that he had not really understood what the word "dogma" meant: but the idea that this mechanism was set in stone lasted for just as long.

Soon, someone had the bright idea of making artificial DNA using the four bases mixed in different proportions, with the other bits of the protein-building machinery, some enzymes and a supply of raw materials. Amazing! A protein was made in the test tube.

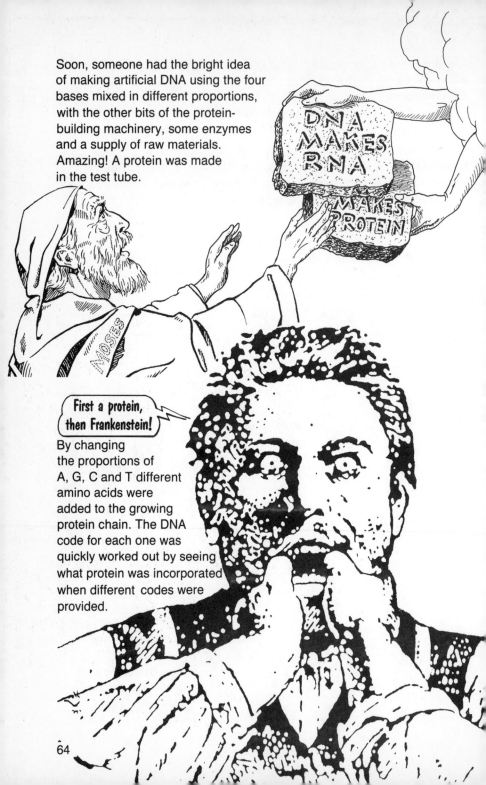

First a protein, then Frankenstein!

By changing the proportions of A, G, C and T different amino acids were added to the growing protein chain. The DNA code for each one was quickly worked out by seeing what protein was incorporated when different codes were provided.

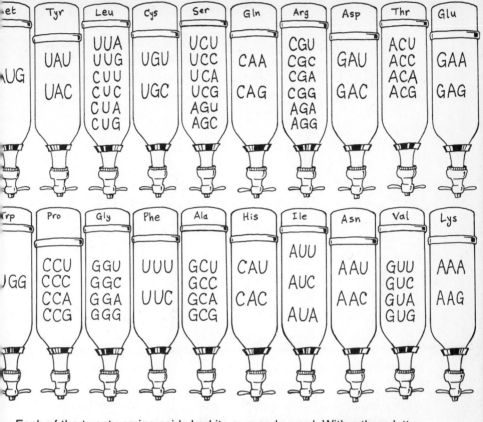

Each of the twenty amino acids had its own code word. With a three-letter word made up of four alternative letters there are 64 combinations. Sometimes several triplets code for the same thing — the code is **redundant**.

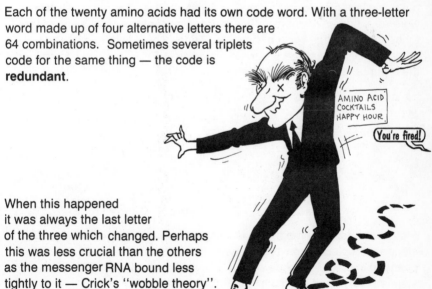

AMINO ACID COCKTAILS HAPPY HOUR

You're fired!

When this happened it was always the last letter of the three which changed. Perhaps this was less crucial than the others as the messenger RNA bound less tightly to it — Crick's "wobble theory".

There was also a three-letter code-word which told the production line when to start, and three more which told it where to stop.

The genetic code was amazingly universal — almost the same from bacteria to humans. Perhaps it stretched back to the dawn of life.

So that seemed to be it: pretty simple, really. Just as Morgan had worked out in his breeding experiments with fruit-flies, the inherited instructions were arranged next to each other in order, and read from one end to the other. The message was **colinear**; the genes were lined up next to each other, the DNA message was read directly into RNA, and that determined the order of amino acids in the proteins.

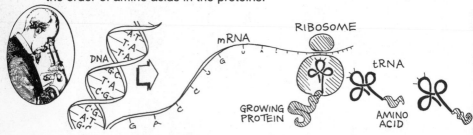

Mutations seemed simple, too. Changing one of the letters in the code stopped the gene from working. Sometimes, a code for an amino acid was changed into a STOP code — and, just as expected, the protein chain stopped growing at that point.

There were still a few details to be worked out, of course.

Bacteria helped again. They had a strange and complicated sex-life with all kinds of ways of exchanging genes. Some even involved "infectious heredity" — viruses carried bacterial genes from one to the other. Perhaps venereal disease had evolved before sex!

One kind of bacterial mating was reasonably conventional: a "male" bacterium started feeding a copy of its DNA into another one, a "female". The process always started in one place on the chromosome and took an hour or so to complete.

A cruel experiment by two French scientists, François Jacob and Jacques Monod, took advantage of this prolonged mating to map the order of the genes.

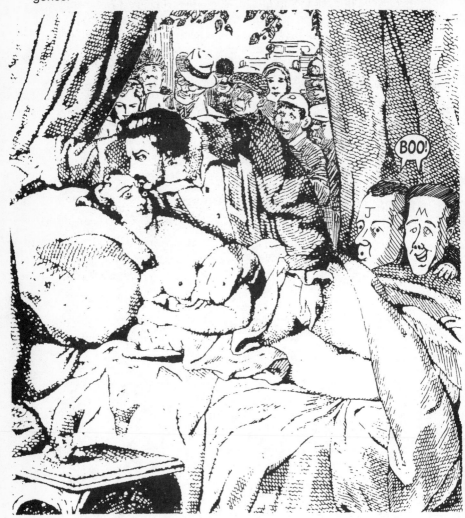

As the bacteria mated, they were suddenly whizzed around in a blending machine. This broke off their relationship — only part of the DNA was inserted. Interrupting mating at intervals after it began gave time for different lengths of DNA to be passed across. Looking at the increasing number of the "male's" genes transferred showed the order in which his genes were arranged — a new way of mapping the DNA.

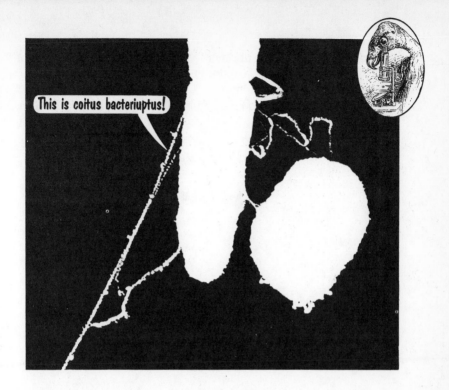

The bacterial DNA map looked quite like Sturtevant's fruit-fly map based on crossing experiments — the genes were arranged in order, one next to the other.

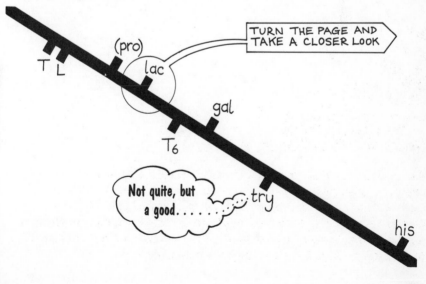

TURN THE PAGE AND TAKE A CLOSER LOOK

Something else soon became obvious — in bacteria, genes which did the same kind of thing were close together. Each group — an "operon" — made a single messenger RNA molecule, which coded for a functioning set of proteins. Things seemed nice and linear here, too.

THE lac (LACTOSE) OPERON

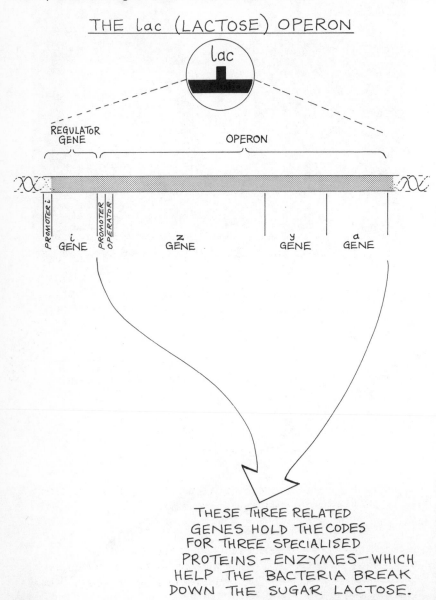

THESE THREE RELATED
GENES HOLD THE CODES
FOR THREE SPECIALISED
PROTEINS — ENZYMES — WHICH
HELP THE BACTERIA BREAK
DOWN THE SUGAR LACTOSE.

But there was a surprise among the bugs.

Bacterial genes were not arranged in a straight line — their chromosome was circular! The genes were arranged as a ring.

Circular chromosomes began to turn up all over the place.

Many cells have lots of DNA outside the nucleus. Most is in the mitochondria, small elements which generate energy from food. Plants have yet more in the chloroplasts, which capture the sun's rays — and make the plant green.

Crossing experiments using mutations in mitochondria and chloroplasts showed that these genes too had an odd pattern of inheritance.

Just as Sturtevant had found in Drosophila, they seemed to be in some kind of order. But making the map proved unexpectedly difficult - the arrangement of the genes seemed to change from experiment to experiment.

In 1954, Ruth Sager had a sudden insight. Cutting a circle at different places changes the patterns of organisation of genes.

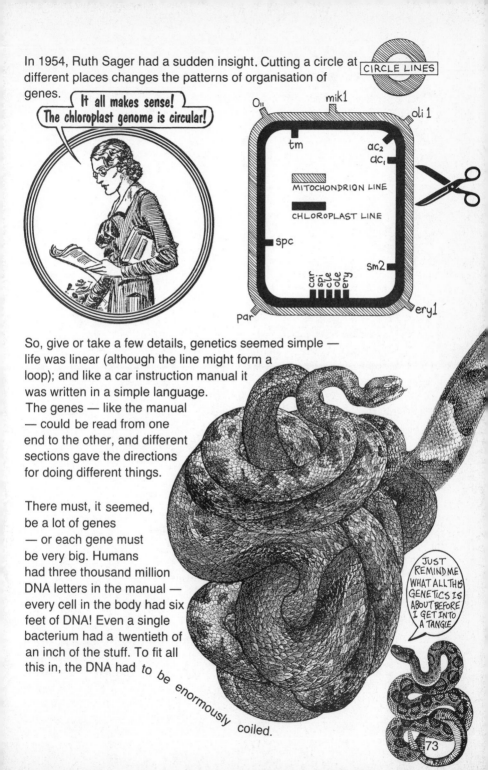

It all makes sense! The chloroplast genome is circular!

CIRCLE LINES

MITOCHONDRION LINE

CHLOROPLAST LINE

O₁₁   mik1   oli 1
tm   ac₂   ac₁
spc
car spi cle ole fray
par   sm2   ery1

So, give or take a few details, genetics seemed simple — life was linear (although the line might form a loop); and like a car instruction manual it was written in a simple language. The genes — like the manual — could be read from one end to the other, and different sections gave the directions for doing different things.

There must, it seemed, be a lot of genes — or each gene must be very big. Humans had three thousand million DNA letters in the manual — every cell in the body had six feet of DNA! Even a single bacterium had a twentieth of an inch of the stuff. To fit all this in, the DNA had to be enormously coiled.

JUST REMIND ME WHAT ALL THIS GENETICS IS ABOUT BEFORE I GET INTO A TANGLE

73

WE INTERRUPT THIS STORY TO BRING YOU A ROUND-UP OF THE BIOLOGY SO FAR! NEARLY ALL LIVING THINGS ARE COMPOSED OF HIGHLY ORGANIZED STRUCTURES CALLED **CELLS**: THERE ARE TWO MAIN TYPES.

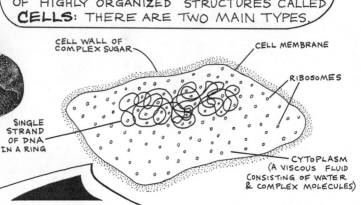

CELL WALL OF COMPLEX SUGAR

CELL MEMBRANE

RIBOSOMES

SINGLE STRAND OF DNA IN A RING

CYTOPLASM (A VISCOUS FLUID CONSISTING OF WATER & COMPLEX MOLECULES)

SMALL & SIMPLE, THE CELL OF A **PROKARYOTE** (SUCH AS A BACTERIUM) HAS NO NUCLEUS & REPRODUCES PRIMARILY BY EQUAL DIVISION.

A THOUSAND TIMES BIGGER (BY VOLUME), THE 'TYPICAL' CELL FROM A **EUKARYOTE** (COMPLEX ORGANISM SUCH AS AN ANIMAL) HAS A NUCLEUS & A MORE COMPLICATED STRUCTURE.

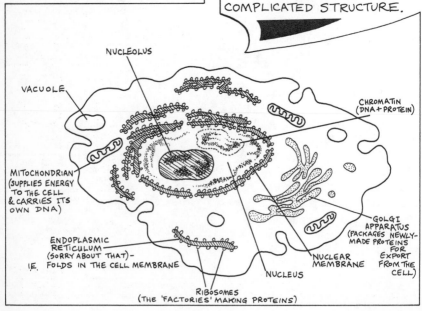

NUCLEOLUS

VACUOLE

CHROMATIN (DNA + PROTEIN)

MITOCHONDRIAN (SUPPLIES ENERGY TO THE CELL & CARRIES ITS OWN DNA)

ENDOPLASMIC RETICULUM (SORRY ABOUT THAT) - I.E. FOLDS IN THE CELL MEMBRANE

GOLGI APPARATUS (PACKAGES NEWLY-MADE PROTEINS FOR EXPORT FROM THE CELL)

NUCLEAR MEMBRANE

NUCLEUS

RIBOSOMES (THE 'FACTORIES' MAKING PROTEINS)

ALMOST ALL THE CELLS IN AN ORGANISM CARRY A COMPLETE COPY OF THAT ORGANISM'S DNA — THE EXCEPTIONS ARE...

... SEX CELLS, THE FEMALE

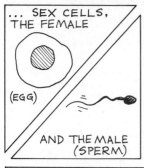

(EGG)

AND THE MALE (SPERM)

EACH CARRIES A HALF-SET OF CHROMOSOMES...

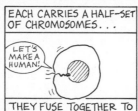

LET'S MAKE A HUMAN!

THEY FUSE TOGETHER TO MAKE A FULL SET (46 IN A HUMAN BEING) & BEGIN THE DEVELOPMENT OF A UNIQUE NEW INDIVIDUAL

THE FERTILIZED EGG DIVIDES...

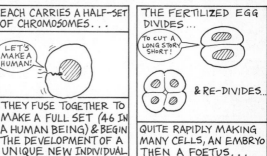

TO CUT A LONG STORY SHORT!

& RE-DIVIDES...

QUITE RAPIDLY MAKING MANY CELLS, AN EMBRYO THEN A FOETUS...

CELLS EVENTUALLY DIFFERENTIATE INTO SPECIALISED SHAPES & FUNCTIONS IN THE ORGANISM.

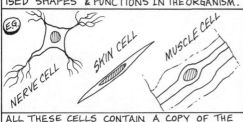

E.G.

NERVE CELL

SKIN CELL

MUSCLE CELL

ALL THESE CELLS CONTAIN A COPY OF THE ORIGINAL SET OF CHROMOSOMES.

BACK TO A 'TYPICAL' HUMAN CELL: INSIDE ITS NUCLEUS ARE 46 LONG INVISIBLE STRANDS OF DNA (THE CHROMOSOMES).

WHEN THE CELL PREPARES TO DIVIDE THE CHROMOSOMES REPLICATE (COPY) THEMSELVES.

THE ORIGINAL & COPY OF EACH REMAIN JOINED TOGETHER

ORIGINAL   COPY

CENTROMERE (JOIN)

THEN THEY THICKEN AND SHORTEN INTO ROD-LIKE SHAPES (VISIBLE UNDER THE MICROSCOPE).

THE MEMBRANE AROUND THE NUCLEUS DISSOLVES...

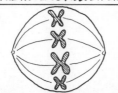

... AND A SPINDLE FORMS ON WHICH THE CHROMOSOMES & COPIES LINE UP.

(FOR SIMPLICITY ONLY FOUR CHROMOSOMES ARE SHOWN)

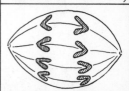

THEN THE ORIGINALS AND COPIES ARE PULLED APART BY THE SPINDLE FIBRES.

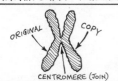

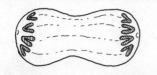

THE CHROMOSOMES ARRIVE AT OPPOSITE POLES AND THE SPINDLE DISPERSES

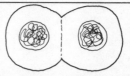

MEMBRANES RE-FORM ROUND 2 NUCLEI, CHROMOSOMES UNWIND & THE CELL DIVIDES.

AND THE SEX CELLS?

HEREDITY

THE 46 HUMAN CHROMOSOMES CAN BE GROUPED IN 23 HOMOLOGOUS (I.E. SAME SHAPE) PAIRS — HERE'S A TYPICAL FEMALE SET...

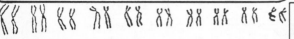

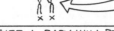

IT MUST BE A FEMALE SET BECAUSE THE FINAL PAIR HAS TWO **X** CHROMOSOMES. ONE OF THESE WOULD BE REPLACED BY A **Y** CHROMOSOME - **X** - IN A MALE SET. THE **Y** CONFERS MALENESS! THE OTHER 22 PAIRS ARE THE SAME.

SO, WHAT DETERMINES WHETHER A BABY WILL BE A BOY OR A GIRL, HAVE BLUE OR BROWN EYES, DARK OR FAIR HAIR? THE SEX-CELLS, THAT'S WHO!

THERE IS A SPECIAL TYPE OF DOUBLE DIVISION FOR MAKING THE SEX-CELLS	RANDOM CHROMOSOME SEGMENTS 'CROSS OVER' BETWEEN PAIRS TO SHUFFLE THE GENES!	FIRST, THE **PAIRS** ARE PULLED APART. WHEN THEY REACH THE POLES...
BEFORE DIVISION, HOMOLOGOUS CHROMOSOMES PAIR OFF (ONLY 3 PAIRS ARE SHOWN HERE),	THE GROUPS OF 4 LINE UP ON THE FIBROUS SPINDLE (AS BEFORE).	...THE SPINDLE DISPERSES AND 2 NEW SPINDLES FORM AT RIGHT-ANGLES.

FOUR CELLS EACH WITH **HALF** THE CHROMOSOMES OF THE ORIGINAL CELL.

THE TWO SETS OF CHROMOSOMES LINE UP ON EACH SPINDLE	THEY SEPARATE, UNWIND & 4 NEW NUCLEI FORM THUS CREATING...	WHICH COPY GOES INTO WHICH CELL IS COMPLETELY RANDOM - THEY ARE INDEPENDENTLY ASSORTED.

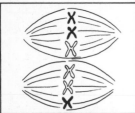

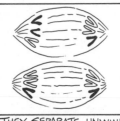

THE SEX OF OFFSPRING IS EQUALLY RANDOM: THE SPERM CARRIES AN **X** & A **Y**, THE EGG CARRIES TWO **X**'s. WHEN THEY MEET THERE IS A ROUGHLY EQUAL CHANCE OF A BOY OR A GIRL BEING BORN.

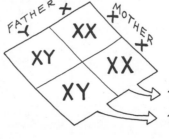

FATHER **X** MOTHER **X**

XY   XX

XY   XX

½ GIRLS
½ BOYS

MEANWHILE, INSIDE THE NUCLEUS...

76

THE MARVEL OF DNA IS ITS ABILITY NOT ONLY TO GOVERN & REGULATE THE PROCESSES INSIDE THE CELL, BUT ALSO TO BUILD ALL THE MACHINERY AND RAW MATERIALS, TOO!

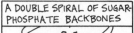

A CHROMOSOME IS A LONG STRAND OF TIGHTLY-WOUND DNA CARRYING GENES ALONG ITS LENGTH—EACH GENE HAS THE INFORMATION TO PRODUCE A FUNCTIONAL PRODUCT.

A DOUBLE SPIRAL OF SUGAR-PHOSPHATE BACKBONES

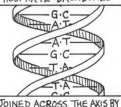

G·C
A·T
A·T
G·C
T·A
T·A

JOINED ACROSS THE AXIS BY COMPLEMENTARY BASE-PAIRS (A TO T & C TO G)

GENETIC INFORMATION IS ENCODED IN THE BASES' FOUR-LETTER LANGUAGE

DNA FLATTENED OUT

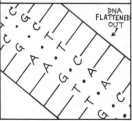

THE CODE ON ONE STRAND IS TRANSCRIBED BY AN ENZYME CAUSING A COMPLEMENTARY MOLECULE TO BE MADE.

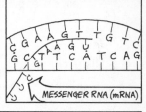

C G A A G T T G T C
G C T T C A T C A G

MESSENGER RNA (mRNA)

mRNA TRAVELS TO A RIBOSOME—A DOUBLE BALL OF PROTEINS & RNA—TO TRANSLATE ITS CODE . . .

U U C G A A G U A G

. . . TO MAKE A PROTEIN!!

EACH TRIPLET OF BASES ON mRNA CALLS UP A COMPLEMENTARY TRIPLET ON THE HEAD OF TRANSFER RNA (tRNA)

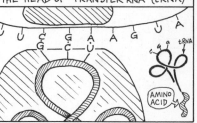

U U C G A A G U A
G — C — U

tRNA

AMINO ACID

EACH tRNA CARRIES ONE OF 20 AMINO ACIDS TO THE RIBOSOME WHICH DETACH FROM tRNA & LINK TOGETHER

← PROTEIN

THE GROWING CHAIN OF AMINO ACIDS TWISTS & FOLDS INTO A **PROTEIN** (MANY PROTEINS ARE ENZYMES)

THE mRNA MOVES ON TO OTHER RIBOSOMES & THE OTHER ELEMENTS DISPERSE UNTIL NEEDED AGAIN,

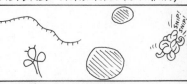

SNIP! SNIP!

WHILE THE ENZYME GOES OFF TO CATALYSE CRUCIAL BIOCHEMICAL REACTIONS IN THE CELL.

. . . THAT'S THE BASICS, NOW BACK TO THE PLOT!

The next job was obvious. To find out how the body worked biologists had to read the DNA manual from the beginning to the end. Everyone expected that such **physical maps** — based on the actual order of the DNA bases — would look much like the old linkage maps based on crossing experiments.

Lots more genes would turn up, no doubt. The physical map of humans would be particularly interesting as so little was known about the linkage map.

To make the new map needed technology — and a lot of money. Genetics stopped being a cheap science.

Tragic, he used to be a cheap scientist.

WIFE & HUMAN GENE- MAPPING PROJECT TO SUPPORT

Lots of the techniques came from taking advantage of bacterial sex.

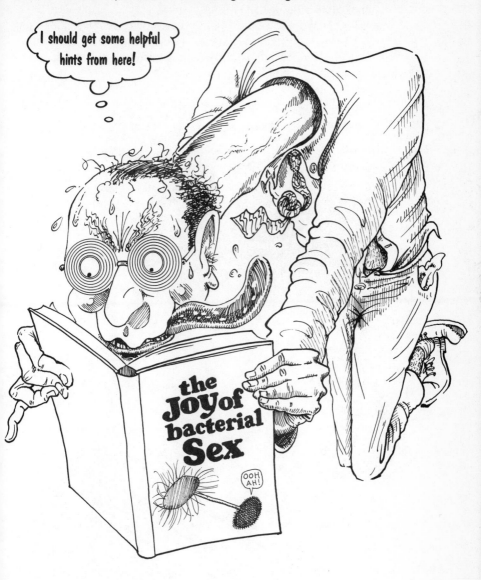

Bacteria do it in lots of different ways. Sometimes they even use viruses to help them to transfer genes — which is, after all, what sex is.

This strange bacterial habit turned out to be very
important in molecular genetics.

When bacteria are infected with viral DNA, they can cut it out with special
"molecular scissors" called restriction enzymes.

These cut
very specific places
in the DNA, each one finding its own special set of DNA letters. There are
lots of different restriction enzymes, each one cutting in a specific place.

They can cut human DNA, too. What's more, certain viruses — or small pieces of bacterial DNA called **plasmids** — can be persuaded to take up the cut pieces. Then they move them into the bacterium, which treats the foreign DNA as if it were its own.

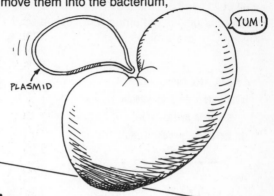

YUM!

PLASMID

engine. It is not necessary to detach the flexible hose. Undo and remove two nuts and bolts securing the exhaust pipe clamp at the manifold to downpipe. Detach the downpipe from the manifold. The speedometer cable should next be detached from the We are to reap the harvest of his son. The broken rancour of your high-swol'n hearts. But lately splinter'd, knit and join'd together. Must gently be preserved, cherished and kept. Forthwith from Ludlow, the young prince; Hither to London, to be crowned our transmission casing. Unscrew the knurled nut and withdraw the cable. Check the rear wheels, jack up the front of the car and support on firmly based axle stands located between the front longitudinal members. Working under

The human gene has been **cloned**; and **recombinant DNA** has been made. This time, though, it's a new kind of recombination — a sort of sex between humans and bacteria; gene exchange between totally different creatures. Amazing!

Grow up a bucket of bacteria and you have millions of copies of the gene, ready to be studied — or even used to make something.

Yeast cells can be used too. They are persuaded to take up and multiply even longer pieces of human DNA. The pieces are called YACS (for Yeast Artificial Chromosomes).

An even simpler way of multiplying genes is to miss out sex altogether.

OH

To copy DNA during a natural cell division and sexual reproduction needs a special enzyme, a **polymerase**. This starts multiplying the molecule when it is told where to go by a short piece of matching DNA, a **primer**.

An INGENIOUS TRICK works wonders. A polymerase is taken from a bacterium which lives in hot springs. It keeps working even at high temperatures. By heating and cooling a DNA sequence plus the polymerase and feeding the mixture with the four DNA bases a chain reaction sets in — millions of copies of the original piece of DNA are made. This is the **polymerase chain reaction** (PCR).

Throw in another handful of G's.

Great! Now let's put it in the fridge.

My gym teacher says that this is no better than sex, but it's a lot quicker.

Once lots of very pure DNA has been made by cloning or by PCR the order of the DNA letters can be read off. This needs new technology, too. One way to start is to cut the DNA into lengths using restriction enzymes which cut particular combinations of letters.

The lengths are separated by **electrophoresis** — pulling them through a molecular maze using an electric current. It's a bit like the first day of the sales — the small and agile get through the aisles quicker than the fat and slow. Taking a snapshot soon after the doors open shows how many people of different sizes there are. What's more, the shoppers — or the DNA — can then be separated by size.

Now we have lots of short bits of DNA, the sections can be read slowly and painfully. Take a piece and use it as a blueprint to grow a series of copies, starting at one end and increasing in length letter by letter. Stopping the process each time a letter is added, means that each copy is slightly longer than the one before. They can be separated by electrophoresis.

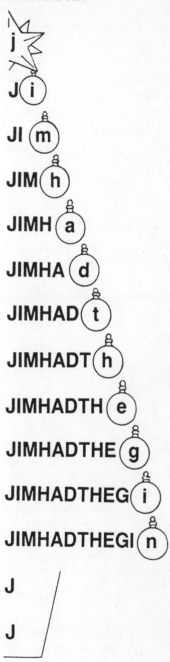

A C G T

ELECTROPHORESIS GEL

The pattern of DNA lengths looks a bit like a Christmas tree. By reading the letters at the ends of the branches the whole message can be read off from end to end.

These and many other tricks mean that the DNA message in any gene can be deciphered.

i
j
J i
JI m
JIM h
JIMH a
JIMHA d
JIMHAD t
JIMHADT h
JIMHADTH e
JIMHADTHE g
JIMHADTHEG i
JIMHADTHEGI n
J
J

In the 1970s, the molecular mappers set to work. Soon, biologists hoped, genetics would be more or less over — the order of all the genes would be sorted out. It would be time to start answering the much more interesting question first asked in Mendel's day. How does a simple inherited message give rise to a complicated thing like a human being (or a pea)?

In 1982 there was the first of a series of nasty shocks. Inheritance in peas, fruit-flies and even humans might LOOK pretty simple when making crosses or looking at pedigrees, but under the simplicity there seemed to be CHAOS in the DNA.

The first genes to be looked at were those which make the red blood pigment. They were ideal — they made large amounts of a pure protein, haemoglobin. The protein is made of two different chains of amino acids. The embryo has a slightly different form, and a related protein — myoglobin — is found in the muscles. What's more, many inherited diseases were already known to be due to faults in red blood pigment. Sickle cell anaemia, the most widespread genetic disease in the world, was due to a change in one of the protein's building blocks.

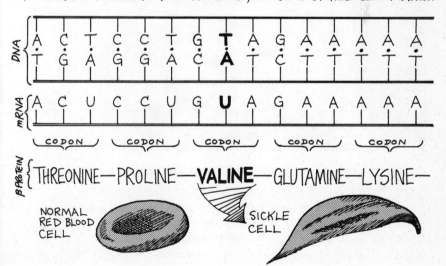

DNA
A C T C C T G A A G A A A A A
T G A G G A C T T C T T T T T

mRNA
A C U C C U G A A G A A A A A

CODON CODON CODON CODON CODON

B PROTEIN {
—THREONINE — PROLINE — GLUTAMINE — GLUTAMINE — LYSINE —

CHAIN OF AMINO ACIDS FORMING NORMAL HAEMOGLOBIN

ONLY **ONE** BASE CHANGE CAUSES A DIFFERENT CODON, WHICH CAUSES A DIFFERENT AMINO ACID TO BE ADDED, CAUSING SICKLE-CELL MUTATION:

DNA
A C T C C T G **T** A G A A A A A
T G A G G A C **A** T C T T T T T

mRNA
A C U C C U G **U** A G A A A A A

CODON CODON CODON CODON CODON

B PROTEIN {
THREONINE — PROLINE — **VALINE** — GLUTAMINE — LYSINE —

NORMAL RED BLOOD CELL

SICKLE CELL

87

Soon, everyone was getting into red blood cells.

Almost at once, there were some interesting findings.

Some made sense. The genes which made the two parts of the haemoglobin molecule were in different places. Each was a member of a family of similar genes close to each other which cooperated to make related things — and were arranged in the order needed during development, embryonic haemoglobin first, adult last, myoglobin not far away.

There were some unexpected, but not particularly alarming, discoveries, too. For example, some of the family members looked a bit like the rest, but did nothing useful. Long ago, a mutation had changed one of their code words to the STOP message. The gene was decrepit, and full of mutations — a **pseudogene**, a living genetic fossil.

NOW, KIDS, DON'T SAY ANYTHING ABOUT YOUR SISTER'S DECREPITUDE

The various inherited diseases of red blood cells were due to different mutations. Some, like sickle cell, were simple — they changed just one letter in the message. Others were due to the loss of whole paragraphs; and in some the DNA of adjacent genes had become stuck together to make a hybrid protein.

All this fitted pretty well into the idea that genes and proteins were much the same kind of thing on different scales.

But then — the first of the big surprises. Suddenly, a lot of the structure of the haemoglobin and other genes began to make no sense at all. Mendel would have hated it!

Reading along the DNA message showed that there was far more DNA within each gene than was needed to make the protein.

The way the gene worked was even stranger. The whole of the DNA of, for example, the ß-chain was read off into a long messenger RNA, but then — astonishingly — sections of the message were cut out and thrown away. An edited version of the message left the nucleus.

It was as if the owner's manual for an English car was interrupted by sentences in Chinese which had to be snipped out before the instructions could be read properly.

Chock the rear wheels, jack up the front of the car and support on の体格を忠実に再現させることになる。axle stands located between the longitudinal members. Move the selector lever to the それはあたかも、生物体が一連の議会選挙区に分けられて、支持する者を当選させるために、各選挙区から代表団が送られてくるかのようだ。この 'D' position. Undo and remove the bolt securing the transmission control cable retainer to the casing. Undo the two control cable adjustment locking nuts and pull ジェミュールは血流に送りこまれる。それから生殖細胞に再び集まり、親 ダーウィンにとってこの仮説は、獲得形質の遺伝をみごとに説明するものだった。たとえばある生物が自分の努力で手足の筋肉を大きくすると、そthe outer の発達した部分からcables出るジェミュールfrom the transmission casing. The の数がふえ、生殖細胞に多く集まる。control inner cable may したがって、now be disconnected from the valve block detent rod and the park その生物の努力の実りが、自動的に子孫に遺伝するのだ。lockrod. 逆に、もし手や足などの器官を使わないでいると、そこから出るジェミュ Make a ールの数は減少し、その不足もまた子に遺伝する。この説は、進化における偶発的変異の役割を予期していた人々に強い確信をもたせた。そして note ダーウィン自身がこのような説を容認したという事実は、ラマルク説を再 評価させる要因になった。1870年から、ラマルクの説の再検討が本格化し of the electrical cable connections at the starter た。そしてダーウィンが死ぬ少し前の数年のうちに、ものすごい勢いのラinhibitor switch and detach the cables. The front of the car should now be lowered to the ground. The weight of ____ ____ unit must now be taken マルク説復活が起こった。

Only part — sometimes a very small part — of the DNA in any gene in higher animals actually coded for a protein. Sometimes a gene was split into dozens of these **exons** by a series of **introns**, the Chinese characters.

Not only were human genes filled with sections that made no sense even though they were read off into the first part of the machinery but, in addition, there were huge regions — millions of DNA bases long — in between the genes that seemed to code for nothing. They made no messenger RNA at all.

In animals and plants, unlike bacteria, most of the DNA was just a few oases of sense in a desert of nonsense. The genetic message of toads and salamanders, in particular, was largely gibberish; but even humans seemed to contain quite a lot of meaningless DNA.

Lots of the molecular desert was — just like the Sahara — very repetitive. Sometimes the same brief message was repeated thousands of times. Often, the order of the DNA letters made a palindrome: they read the same backwards as forwards.

the broken rancour of your high-swol'n hearts, but lately splinter'd, knit and join'd together, Must gently be preserved, cherished and kept. Forthwith from Ludlow, retsina canister retsina canister retsina canister retsina canister retsina canister retsina canister retsina canister retsina canister able was I ere I saw Elba able was I ere I saw Elba able was I ere I saw Elba able was I ere I saw Elba Detartrated Detartrated Detartrated Detartrated Detartrated Madam I'm Adam Madam I'm Adam Madam I'm Adam Madam I'm Adam Madam I'm Adam Madam I'm Adam Madam I'm Adam Madam I'm Adam Madam I'm Adam Madam I'm Adam A man, a plan, a canal, Panama! A man, a plan, a canal, Panama! A man, a plan, a canal, Panama A man, a plan, a canal, Panama! A man, a plan, a cana Panama! A man, a plan, a canal, Panama! Glenelg Glenel Glenelg Glenelg Glenelg Glenelg Glenelg Glenelg Glene Glenelg Glenelg Glenelg Glenelg Glenelg Glenelg Glene Glenelg Glenelg Glenelg Glenelg Glenelg Glenelg Glen Glenelg Glenelg Glenelg Glenelg Glenelg Glenelg Glen Glenelg Glenelg Glenelg Glenelg Glenelg Glenelg Gler Glenelg Glenelg Glenelg Glenelg Glenelg Glenelg Gler Glenelg Glenelg Glenelg Glenelg Glenelg Glenelg Gle Glenelg Glenelg Glenelg Glenelg Glenelg Glenelg Gle Glenelg Glenelg Glenelg Glenelg Glenelg Glenelg Gle Glenelg Glenelg Glenelg Glenelg Glenelg Glenelg Gle Glenelg Glenelg Glenelg Glenelg Glenelg Glenelg Gl Glenelg Glenelg Glenelg Glenelg Rotavator Rotavator Rota Rotavator Rotavator Rotavator Rotavator Rotavator Rota Rotavator the young prince; hither to London. crowned our

A few of these "tandem repeats" were scattered throughout the DNA. Different people had different numbers of copies of the repeated word dispersed in different places. Cutting their DNA wherever that combination of letters was found gave a unique mix of lengths of DNA — a "genetic fingerprint".

Where did all this spare DNA come from? It certainly didn't seem to **do** much.

There was an extraordinary hint in the strange circular genomes of bacteria and mitochondria. Perhaps lots of our DNA was once parasitic!

Mitochondria and bacteria were alike in another way — a poison which kills bacteria does the same to mitochondria, but leaves the rest of the cell alone.

Perhaps mitochondria once *were* bacteria! Long ago, they invaded cells and lived as parasites inside until they found a useful role. If so, a lot of the DNA of the higher organisms once behaved in its own interests rather than that of its host!

Even nuclear genes may behave in a selfish way. Some wild mice have a mutation which shortens the tail. Two doses are lethal — but in spite of this disadvantage, in some places the short tail gene is common.

When a male with a single copy makes sperm, the half which carry the mutated gene cheat: they fertilize many more than half the eggs so that their harmful gene spreads in the next generation. Females do not like this a bit and make every effort not to mate with a male carrying the selfish gene.

By behaving in this unfair way the short-tail gene spreads through the population even though it damages many of its descendants.

Cut off their tails with a carving knife!

Soon, there was a nasty suspicion that a lot of DNA was like this: perhaps an awful lot did its carriers no particular good, even if it usually did not do too much harm.

Genes were beginning to look less pure and less simple than Watson and Crick had thought.

As well as there being far more DNA than was needed, different creatures had very different amounts, and in a way that made little sense.

Humans have about a thousand times more DNA than bacteria, it's true; but we salamanders have about twenty times more than humans!

Does this suggest that we are twenty times more complicated than humans?

What *is* going on?

There was a hint among the toads and salamanders. Nearly all their vast excess of DNA was made up of repeats of the same message.

Sometimes, two closely-related toad species had millions of copies of a different DNA repeated sequence in each. Although they looked almost the same most of their DNA was completely different! Perhaps it was evolving in its own interests, and not in that of their carriers, and had spread through each species since their evolutionary split.

All these DNA-rich creatures had evolved slowly, scarcely changing for millions of years. The slowest of all were the lungfish — "living fossils", who look much like the first land vertebrates. Today, their cells are swollen with huge amounts of DNA. But their fossils show that, long ago, when they evolved quickly as they first came ashore, the cells — and the amount of DNA — were of normal size. Perhaps the invading DNA had actually slowed down their evolution.

Could most of the DNA, even human DNA, be selfish; acting in its own interests and taking over its host, given a chance?

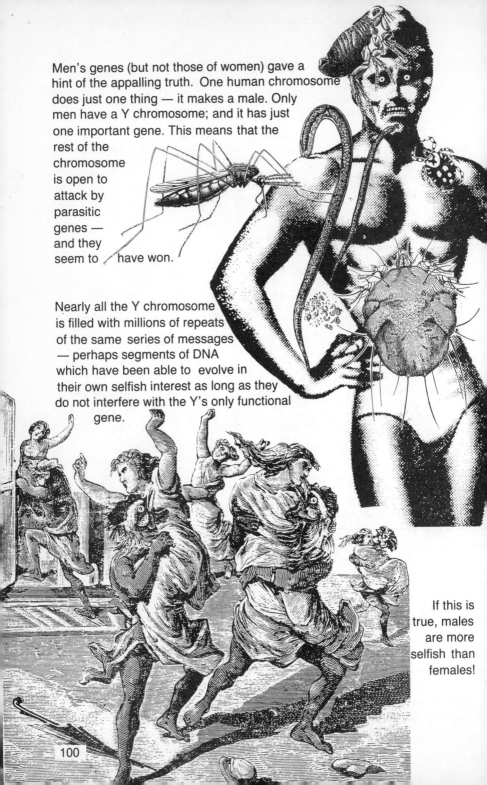

Men's genes (but not those of women) gave a hint of the appalling truth. One human chromosome does just one thing — it makes a male. Only men have a Y chromosome; and it has just one important gene. This means that the rest of the chromosome is open to attack by parasitic genes — and they seem to have won.

Nearly all the Y chromosome is filled with millions of repeats of the same series of messages — perhaps segments of DNA which have been able to evolve in their own selfish interest as long as they do not interfere with the Y's only functional gene.

If this is true, males are more selfish than females!

Parasitic DNA even causes mutations. Once, mutations seemed straightforward — like darts hitting targets. Even forty years ago, there was a nasty suspicion that things were not so simple. In the 1950s, Barbara McClintock was working on mutations in maize.

These are very useful things: genetic changes as the seeds develop can change the colour of patches of cells from yellow to black.

A few mutations at the colour gene turned up now and again, as might be expected. But if another gene was crossed into the maize, the mutation rate from yellow to black rocketed up! One gene seemed to be causing mutations in another. Even more baffling, crossing experiments showed that the position of the mutator gene on the chromosome shifted from generation to generation.

It looked as if it could move around, causing damage wherever it landed. In fact, it is a piece of mobile DNA which shifts camp now and again. Usually, it causes no trouble, but now and again it settles down in a place which damages its host.

Now, all kinds of mutations in various creatures seem to be due to the same thing. Many of the visible mutations used in Drosophila genetics are due to the insertion of a piece of migratory DNA into functional genes.

At least one human disease, neurofibromatosis, is due to the insertion of an extra piece of DNA into a working gene. Often, the symptoms are mild and scarcely noticeable, but sometimes they are severe. Some believe that Joseph Merrick, the Elephant Man, had this inherited abnormality in an unusually damaging form.

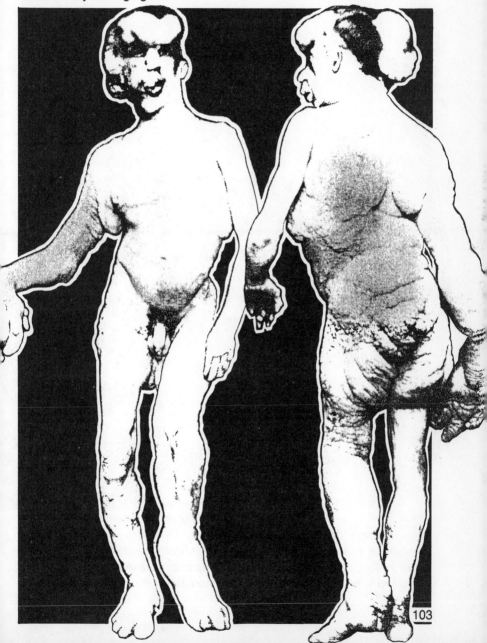

The molecular parasites multiply quickly, given the chance. One type —
about 3000 DNA letters long — has invaded the fruit-flies Morgan worked
on since his day. The stocks which he and his students collected fifty
years ago are free of these "P-elements"...

... but modern flies from the same
place can have dozens of copies of this short DNA sequence inserted into
their genes — and many of them cause mutations, given the chance! They
seem to have moved into the laboratory species of **Drosophila** from an
unrelated species that lives in the jungles of South America.

Even humans are not safe from having their DNA constantly mixed up. The bits of DNA which make up the "genetic fingerprint" move around the genome quite often, although they do not do any harm. In some inherited diseases, there is worse news.

"Fragile-X" is the commonest cause of inborn mental defect. It is due to the insertion of a piece of mobile DNA into the X chromosome. Comparing parents and children shows that each generation, the number of copies goes up; and the disease gets more damaging as children succeed their parents.

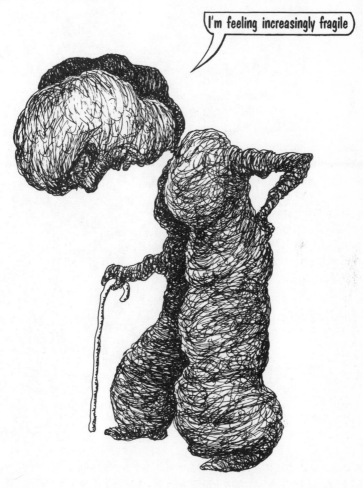

I'm feeling increasingly fragile

Mendel's hereditary particles are beginning to look rather fluid!

But still, it seemed, the basics were right. At least genes were a **bit** like an instruction manual. Although lots of nonsense may have crept in, genes were in some kind of order and could be read from one end to the other.

In the 1990s, new horrors began to emerge.

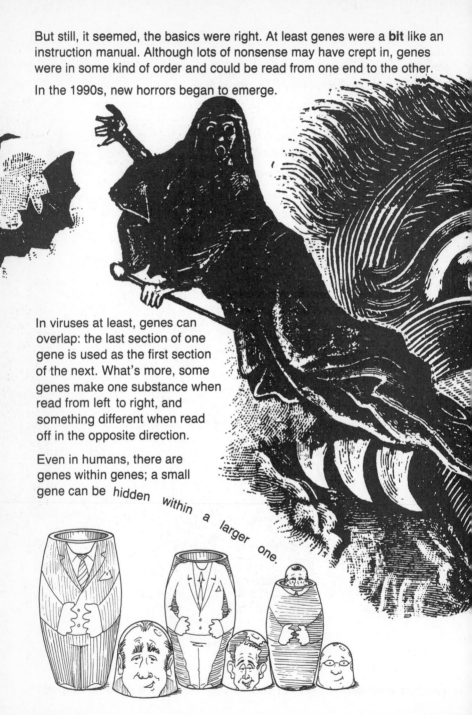

In viruses at least, genes can overlap: the last section of one gene is used as the first section of the next. What's more, some genes make one substance when read from left to right, and something different when read off in the opposite direction.

Even in humans, there are genes within genes; a small gene can be *hidden within a larger one.*

Even Watson's Grand Central Dogma was not safe!

Some viruses — the one which causes influenza, for example — use RNA, not DNA, as their genetic material. The RNA contains their protein-coding machinery.

Geneticists did not find this too alarming. Many felt that perhaps RNA was the original genetic material at the beginning of life three thousand million years ago. Like DNA, it contains information written in a four-letter code; and, unlike DNA, it can copy itself without the help of enzymes.

Perhaps the viruses were frozen accidents from the distant evolutionary past.

Much more baffling, though, was the discovery that some very tiny viruses seem to have no nucleic acids at all: their genetic information might be coded directly into proteins!

They include the virus for a sheep brain disease called scrapie, and for a similar disease in humans — once transmitted by cannibalism in Papua New Guinea.

The new particles were named Prions (for Protein Virions).

The name led to a bizarre misunderstanding.

Prions are in fact antarctic sea-birds related to albatrosses.

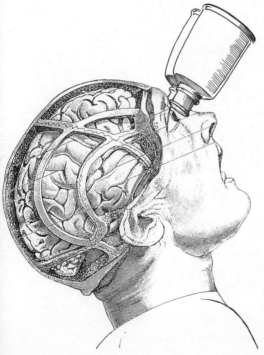

Could, an ornithologist asked in a letter to *Nature* soon after the discovery was announced, such large birds really be the agents of an infectious brain disease? His mind boggled.

Geneticists were equally surprised by the finding that, sometimes, the flow of genetic information changed direction. RNA could make DNA — the Central Dogma in reverse.

The RNA-based viruses (**retroviruses**, as they are known) have a special enzyme, a **reverse transcriptase**, which copies their own message into a DNA segment. This is inserted into their host's DNA, which is forced to make many copies of the RNA retrovirus.

Retroviruses are important as some of them can give rise to cancer (sometimes by picking up and modifying human genes and returning them to the DNA). AIDS is due to infection by a retrovirus known as a Human Immunodeficiency Virus (HIV). It infects white blood cells and suppresses the immune response, so that the body is prey to devastating infections.

In spite of the new confusion in genetics, it is clear that — just as at the first explorations of South America — hidden within the genetic map there are new and startling facts about genes, about disease and about evolution. Now there is a scheme for the great Map of Ourselves — a list of all the three thousand million letters in the human DNA. It may be complete by the year 2000.

It will cost a lot: but like many maps will be the first step to exploiting the country which it reveals.

If nothing else, it will be the largest atlas ever published.

The map contains all the information about our genes. The size of a typical single gene in the whole length of DNA is that of an ant compared to Mount Everest: and the job of finding it is not much easier.

The best way to start is to look at a gene which has gone wrong.

There are plenty of these — and they are getting more important as infectious disease is controlled.

Most young children in hospital are there because of inborn disease. If all diseases (such as heart disease and cancer) which have some inherited component are included, then most people die for reasons to do with their genes.

Although there is a lot of inborn disease about, most individual diseases with a simple pattern of inheritance are rare. Around six thousand different ones are known — some of them very rare indeed.

We're the last generation to die of infection rather than inborn disease.

Some which are rare in most places are common in others — perhaps because many of the people living there descend from a single ancestor with the gene. For example, the gene for the nervous illness Tay-Sachs Disease is frequent among Ashkenazi Jews.

Afrikaners have a number of inherited diseases which are very rare elsewhere, as all of them descend from just a few founders some of whom, by chance, carried these genes. In West Africa and elsewhere sickle-cell and other haemoglobin errors are common because those with one copy of the gene are protected against malaria.

What can genetics do about inborn disease? At the moment, not very much — but there are great hopes for the future . . .

In the western world, the commonest single gene defect is Cystic Fibrosis. It was finally tracked down in 1990 — an astonishingly short time after the search was started. Its story is likely to be repeated for many inherited illnesses — and it shows how useful the genetic map is likely to be.

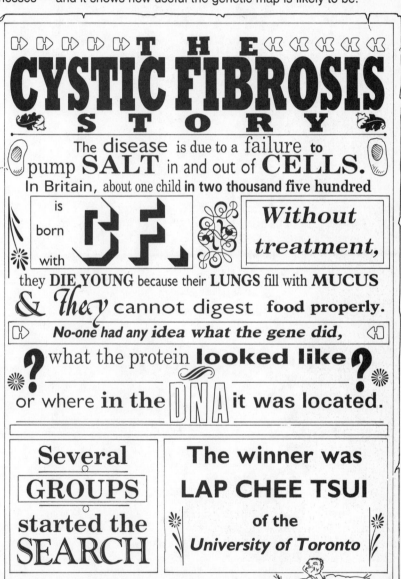

THE CYSTIC FIBROSIS STORY

The disease is due to a failure to pump SALT in and out of CELLS.

In Britain, about one child in two thousand five hundred is born with C.F.

Without treatment, they DIE YOUNG because their LUNGS fill with MUCUS & they cannot digest food properly.

No-one had any idea what the gene did, what the protein looked like or where in the DNA it was located.

Several GROUPS started the SEARCH

The winner was LAP CHEE TSUI of the University of Toronto

The first hint as to where it was came from old-fashioned family studies. The CF gene travelled down the generations in a way that showed that it was not on the sex chromosomes. Soon, it was found to be linked to a change in the sequence of DNA which had already been tracked to a small part of chromosome 7.

This segment was cut out and inserted into mouse cells in the laboratory. Slowly, thousands of letters were read off. Most of the message made no sense at all in the three-letter words of the genetic code.

Now and again, it began to say something. Segments of DNA began to read as if they could produce proteins. What most of these did was quite unknown.

However, one DNA section looked as if it could produce a protein similar to one in the membranes of other creatures.

In families with cystic fibrosis this section of the DNA followed exactly the inheritance of CF.

Now it was possible to reconstruct what the protein must look like from its DNA sequence. This is **reverse genetics** — inferring what a protein looks like, what it does, and what went wrong from the order of DNA letters, rather than using the structure of a damaged protein to work out the changes in DNA.

Once the order of the amino acids in the protein has been worked out from the order of the DNA letters, its shape can be predicted. This might even be the first step to designing a drug which could correct the damage!

Now, dozens of genes — including those for most of the severe inherited diseases — have been tracked down in much the same way. The French have a very successful programme based on television appeals for cash — the Genethon.

Now it's — Hunt the Gene;
Maintenant, chassez la gene!

There are all kinds of new methods of gene mapping, more are appearing all the time. One of the cleverest is to use DNA's ability to bind to copies of itself. To search for where in the DNA a particular protein is coded for, the order of its amino acids is read off. From this, a matching set of the DNA letters based on the three letter code can be made. This is labelled with a fluorescent dye, and poured onto living cells. It sticks onto the chromosomes in exactly the place where the appropriate DNA is.

The method is called FISHing (for Fluorescent In-Situ Hybridisation) for genes.

There's still a hundred thousand or so genes to be found before even the main landmarks in the map have been sorted out. For most there are no convenient inherited diseases to hint at what they might do.

There are also thousands of millions of DNA letters which seem to do nothing — they are not even transcribed.

There is much argument about what to do next.

Is it really worth penetrating into the forest of nonsense, which is what most of the DNA seems to be, or should we stick to the villages and towns where something is actually made?

Good point! One way of getting at those working genes is to hunt just for messenger RNA's — evidence that the gene is producing something — and work back from there to get the DNA sequence.

The brain is a good spot to look — it is a complicated place, and has about thirty thousand genes working at a time. Other cells (such as those in the blood) have far fewer. Five thousand brain genes have already been identified and reading their DNA message is well under way.

I think therefore I amino acid!

However, some biologists believe that it is worth looking for treasure in the depths of the molecular forest — after all, no-one has any idea what could be hidden there.

Their approach is to forget about what the DNA actually does, and to make a crude large-scale map of great sections of chromosome whose details can be filled in later. It's as if a few words per page were read from random sections of the whole of the manual; and the pages put in order by seeing where paragraphs overlap.

the breather hose from the cylinder head cover. Undo and remove the nut and disconnect the accelerator rod from the kick-down bellcrank lever. Undo and remove the four nuts securing the carburettor to the inlet manifold. Lift away the carburettor, accelerator cable abutment bracket and air cleaner, move to one side of the engine compartment.

Remove the carburettor distance pieces, gaskets and heat shield from the inlet manifold. Undo and remove the banjo bolt securing the servo pipe to the inlet manifold. Recover the two copper washers. Move the servo pipe clear of the engine. Undo and remove the two nuts and bolts securing the exhaust pipe clamp halves at the manifold to downpipe connection. Undo and remove the two nuts and release the exhaust pipe from the bracket on the transmission unit.

It is best to mount the engine on a dismantling stand, but if this is not available, stand the engine on a strong bench at a comfortable working height. Failing this, it can be stripped down on the floor. During the dismantling process, the greatest care should be taken to keep the exposed parts free from dirt. As an aid to achieving this thoroughly clean down the outside of the engine, first removing all traces of oil and congealed dirt.

A good grease solvent will make the job much easier for, after the solvent has been applied and allowed to stan for a time, a vigorous jet of water will wash off the solver and the grease with it. If the dirt is thick and deeply embedded work the solvent into it with a strong stiff brush. Finally wipe

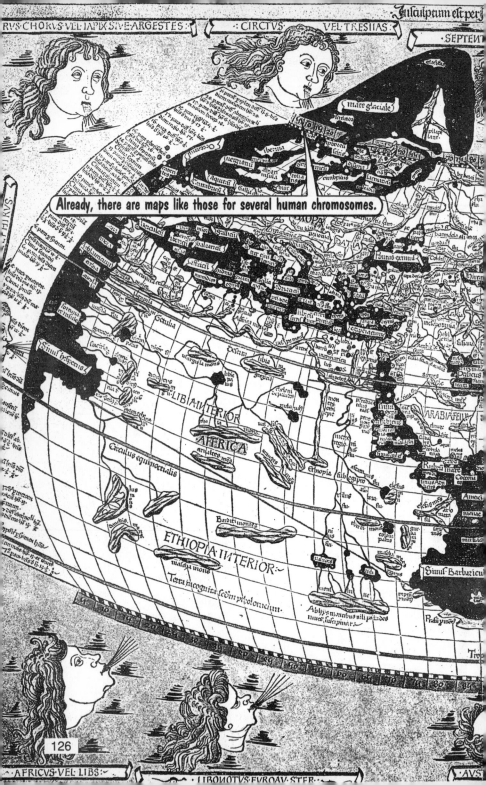

Already, there are maps like those for several human chromosomes.

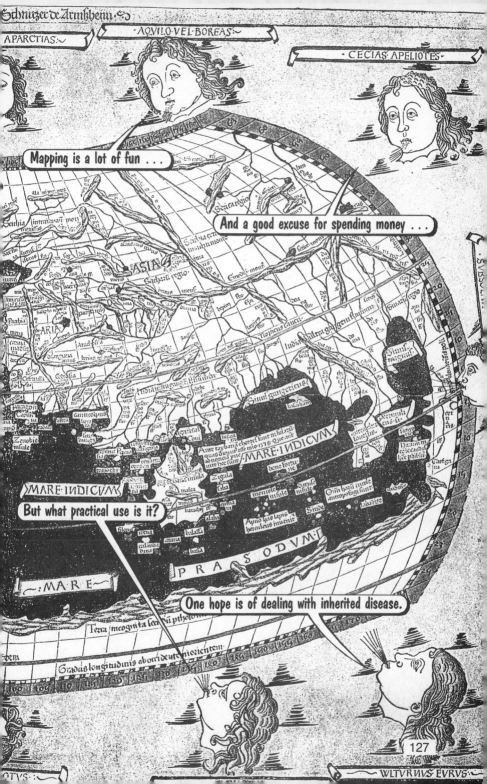

Medicine has always been better at prevention than cure. The triumph over infection was due to sewers more than to antibiotics. Medical treatment came much later.

Genetic disease is no exception. For most, the best that can be offered is not treatment, but diagnosis of an affected embryo.

Who are you calling "affected"?

Parents often choose to end a pregnancy if told that the child will be born with a severe illness. For the most severe diseases, nine out of ten accept genetical advice; and already the number of children born with some inherited disorder is dropping rapidly.

Prevention is better than sewer (particularly *this* one).

But genetics can do more to prevent inherited disease.

For recessives (where two copies are needed to do any harm) it can identify carriers — people who have only one. If two carriers marry, they are at risk of having an affected child.

Sometimes the information helps. In Orthodox Jewish societies, match-makers are still important in helping to arrange marriages. They are told if two potential mates are each carrying a single copy of the recessive gene for Tay-Sachs Disease. Perhaps this will persuade them that they are not ideal partners.

But everything is not always so simple. Take cystic fibrosis. Among white-skinned Britons and Americans, one baby in two thousand five hundred is born with the disease; but about one person in twenty-five is a carrier!

Almost none of the millions of the people involved know anything about it, and the gene does them no harm. In nine out of ten carrier families, no case of the disease has ever been recorded.

To test whether someone is a carrier is straightforward — a simple mouthwash does the job. It costs only about the same as a decent restaurant meal. Three out of four people offered the test take it.

But is there any real point in screening thousands of healthy people, and telling them the truth? Certainly, parents who have had a child with the disease think so: nine out of ten feel that a universal screen should be available so that others do not have to go through their family's experiences. Many of these parents have no more children once a CF child has been born.

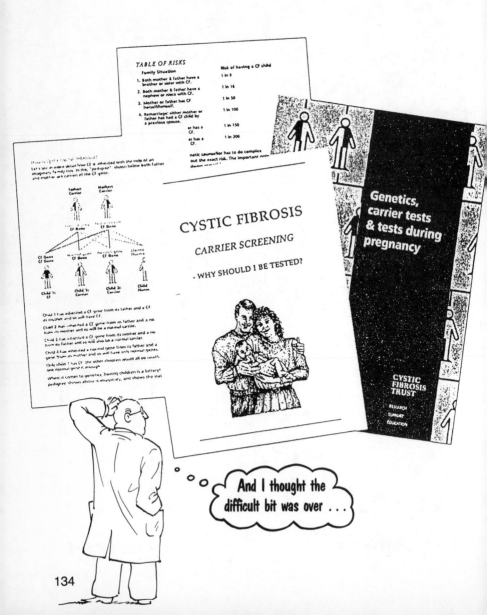

What's more, looking for carriers is a lot more difficult than once hoped. At the molecular level, more than two hundred and fifty different changes can damage the cystic fibrosis gene. Some of the mutations are much more damaging than others.

Even the best test misses some carriers. Even worse, there are big changes from place to place — the British test would miss **most** carriers in Turkey.

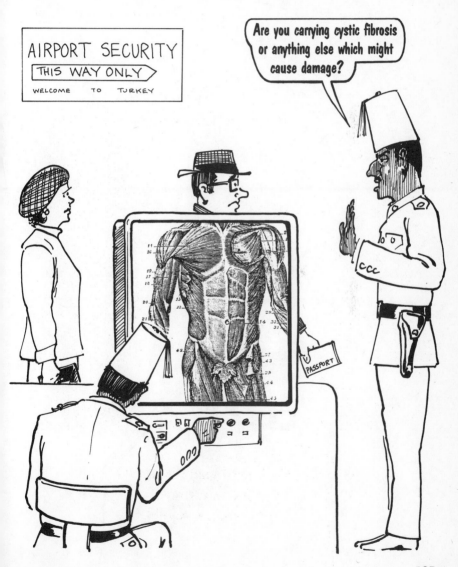

Screening whole populations for a wide range of inherited diseases may never come.

But genetics does have more positive aspects; and making the genetic map may be the first step towards treating, or even curing, disease.

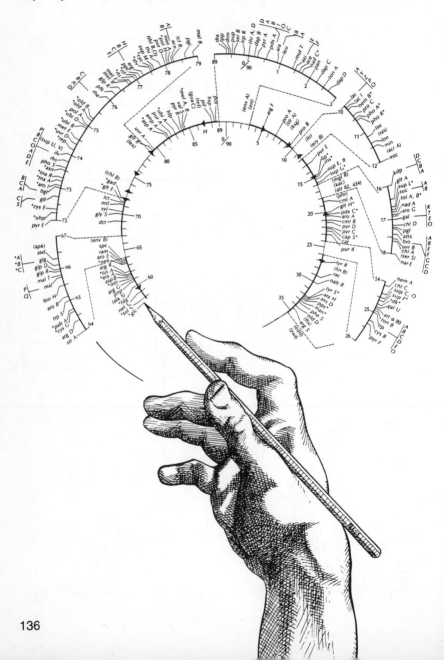

Of course, if a disease is inherited, this does not mean that it cannot be treated. Long before the gene was found, new treatments for cystic fibrosis (such as loosening the mucus in the lungs) allowed children with the disease to survive far better than before — and now that the protein is known, there is new hope for a drug.

A more drastic remedy is to transplant the heart and lungs from a normal person into someone with cystic fibrosis.

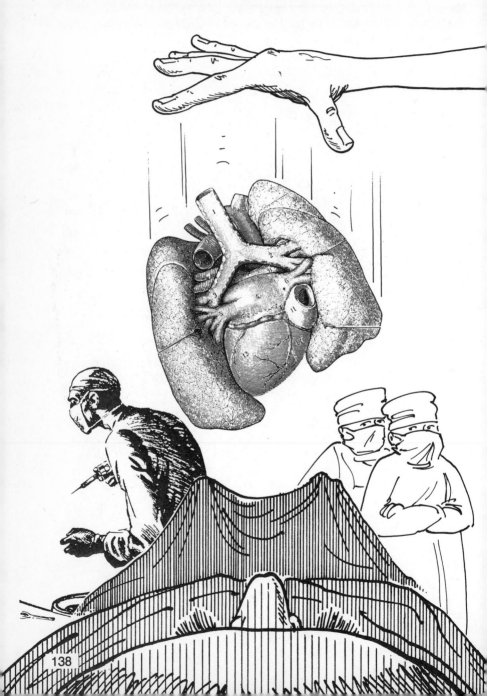

Now, there is the prospect of less dramatic treatments for CF and other inherited diseases. Genes can be moved around from place to place in the living world.

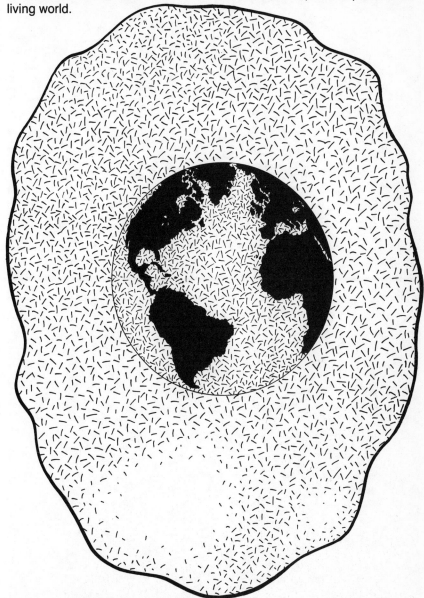

There is a new science of genetic engineering. Perhaps this will improve the treatment of genetic disease.

The engineers have many ways to interfere with genes.

Cells can be persuaded to take up foreign DNA by applying an electrical current which opens up pores in their membrane.

Sometimes, shooting at cells using tiny golden bullets loaded with foreign DNA persuades them to incorporate the alien gene.

The best way to move genes is to use viruses to make the link. Several human genes have been inserted into bacteria; and the bacteria then act as factories to make the gene's product. Insulin and the protein which goes wrong in the blood-clotting disease haemophilia are both made in this way.

Genes can even be put into animals. Engineered sheep produce human growth hormone in their milk: and there is even hope of engineering potatoes to make human proteins which could be used in medicine.

Moo!

Working copies of the gene that has gone wrong in CF can be put into living cells, and these sprayed into the lungs of patients. They help to deal with the symptoms of the illness.

The real breakthrough would be gene therapy for those with a damaged gene: replacing a faulty gene with a working one. This is already commonplace for organs: why not genes?

Now, where are my jeans?

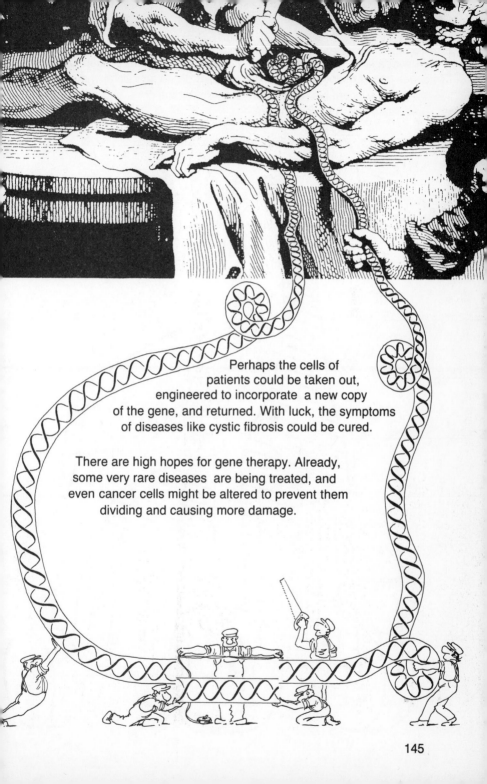

Perhaps the cells of
patients could be taken out,
engineered to incorporate a new copy
of the gene, and returned. With luck, the symptoms
of diseases like cystic fibrosis could be cured.

There are high hopes for gene therapy. Already,
some very rare diseases are being treated, and
even cancer cells might be altered to prevent them
dividing and causing more damage.

One idea is to insert genes that will make the cell more susceptible to drugs into cancer cells. Since cancer cells divide so quickly they take up the 'drug-susceptible' gene more than do normal cells, and the drug kills only those which have gone wrong.

Cancer cells show other changes which can be attacked by the genetical arms industry. Sometimes the clues of identity on the cell surface change when a cell becomes cancerous. Making a matching copy of these new clues and linking drugs to the copy means that the drug can be targeted straight at the cancer cell.

In mice, much more has been done. Genes can be inserted into egg cells so that the engineered DNA is passed to later generations. **Transgenic** mice are important as models of human inherited disease (including cystic fibrosis) which can be used to test drugs

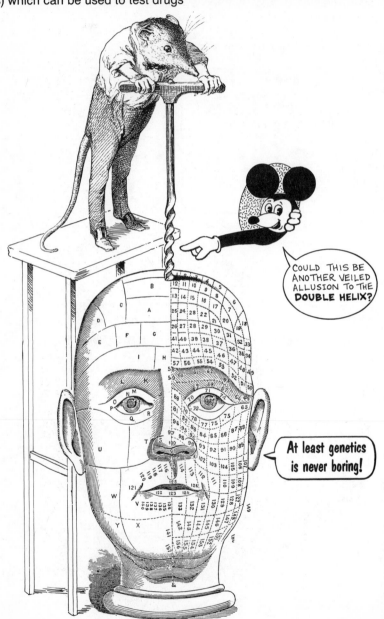

Another clever trick is to interfere with the identity of a useful species by changing the genes which control chemical clues on the cell surface. These **antigens** mean that organs cannot easily be transplanted from one individual to another — let alone from between species. Now, genes for human antigens have been put into pig eggs and transmitted to later generations. Perhaps it will soon be possible to transplant pig organs into humans.

But germ-line therapy, as it is known, is too close to Frankenstein for most geneticists. There are no plans to use it on humans themselves.

As in most of medicine, genetic discoveries bring ethical problems; germline therapy is just one.

Inevitably, diagnosis of inherited disease before birth means that abortion is the only advice which can be offered. In many places, this is a real problem. In the USA, the anti-abortion lobby is so strong that many charities dare not campaign openly for diagnostic tests for disease.

The ethics of abortion are different from those in the rest of medicine; the decision is made on behalf of someone else — the unborn child — rather than by the person actually involved. Some people are even concerned that there will be pressure — from government or health insurers — to terminate genetically damaged foetuses to save the cost of treatment. Privacy about the results of tests is essential.

The law has got involved — in the USA there have been "wrongful birth" law-suits by parents who have had genetically damaged children because they were not diagnosed before birth. Even the children themselves have sued, to get money for treatment. For a disease like cystic fibrosis, where even the best tests cannot pick up all carriers, this is a real problem.

Genetics is running into another, unexpected, problem: it is raising too many hopes and providing too much information. Many people believe that it can do much more than it really can — and seem to be happy to accept treatments which most geneticists would not agree to. The public view of what is right often differs from that of the professionals.

In America, abortion is sometimes hard to get. But three out of four Americans would happily accept germ-line therapy. Some American parents have even asked for growth hormone genes to be inserted into children, so that they grow tall.

Others even suggest that they would accept the idea of inserting genes for increased intelligence — four out of ten Americans believe that this is a good idea! The technique is not remotely feasible, and probably never will be.

Others believe — like Francis Galton himself — that . . .

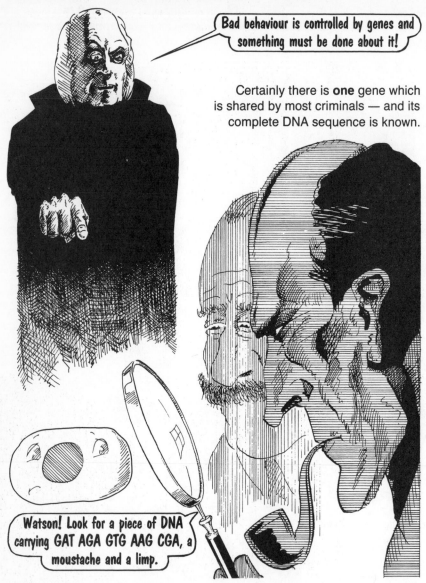

Bad behaviour is controlled by genes and something must be done about it!

Certainly there is **one** gene which is shared by most criminals — and its complete DNA sequence is known.

Watson! Look for a piece of DNA carrying GAT AGA GTG AAG CGA, a moustache and a limp.

It is the single small gene, carried on the Y chromosome, which makes its carriers male. Most criminals are men: the criminality gene has been found! Needless to say, no-one suggests that geneticists should do anything about it or even that this says anything useful about crime.

When it comes to women, though, things are different. It is easy to tell the sex of a foetus by looking at its chromosomes. In Britain, parents do not have a strong preference for boys or girls —

In India we do: and there are clinics which will terminate female foetuses for a fee!

To the alarm of geneticists . . .

Oh no! Our subject is being used to promote prejudice against women!

There is even prejudice against those who carry genes for inherited disease. Many black people in the United States have a single copy of the gene for sickle-cell anaemia. Most know nothing about it, and have no symptoms. A screening programme in the 1970s led to discrimination in jobs, and to much unnecessary misery in those diagnosed as carriers.

BLACKS   WHITES   SICKLE-CELL CARRIERS

However, on the average, everyone carries single copies of two different genes which — just like sickle-cell — would, if present in double dose, kill their carriers. Generally, no-one worries about it — the sickle-cell fiasco was just a matter of bad planning. In Canada, screening for genes like cystic fibrosis is even carried out in school biology lessons.

But sometimes, the information **is** worrying. Some abnormal genes are dominant: they kill when just one copy is present — either at birth, or in early middle age. A test for one of these — Huntington's Disease, a degeneration of the nervous system — is now available. Some people found to be carrying the gene have threatened suicide, In fact, most of those at risk have chosen not to take the test. They prefer uncertainty to knowledge.

Huntington's disease is — fortunately — rare. But other dominant conditions are not. For one — polycystic kidney disease — around 50,000 British people are at risk. They may at any time have kidney failure about which nothing much can be done. A test exists for the gene — but will people really want to know their destiny?

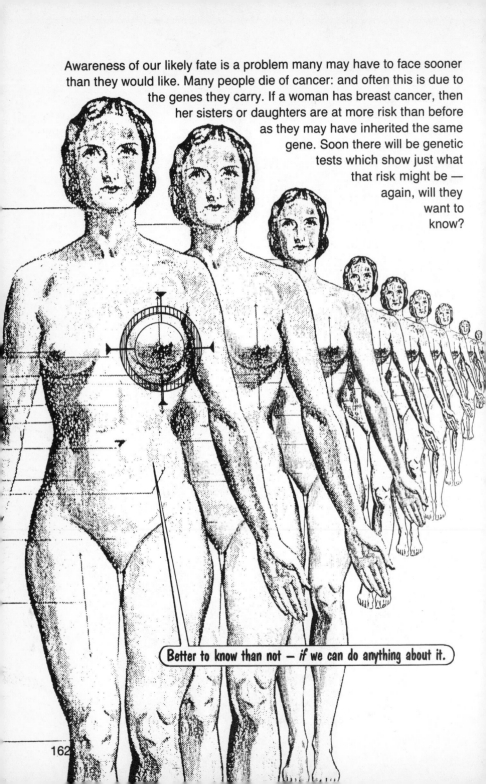

Awareness of our likely fate is a problem many may have to face sooner than they would like. Many people die of cancer: and often this is due to the genes they carry. If a woman has breast cancer, then her sisters or daughters are at more risk than before as they may have inherited the same gene. Soon there will be genetic tests which show just what that risk might be — again, will they want to know?

Better to know than not — *if* we can do anything about it.

For some inherited cancers, like cancer of the bowel, some doctors suggest that those with the gene

should have an operation to remove the bowel even before symptoms appear.

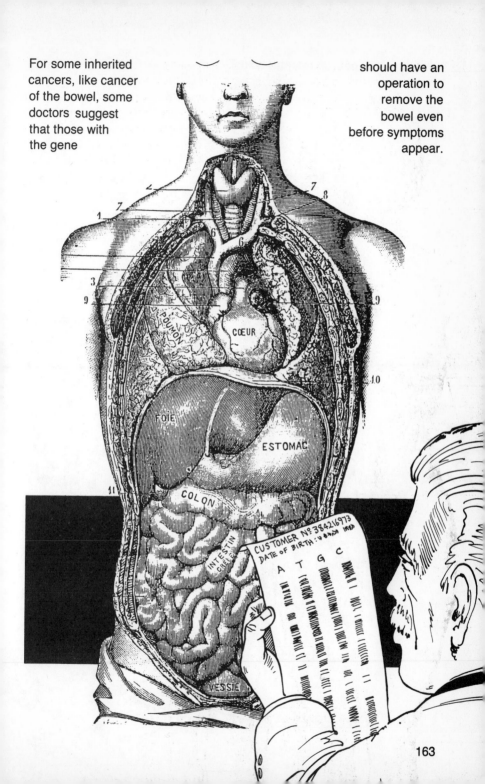

Knowing about genetic risk may actually help:—

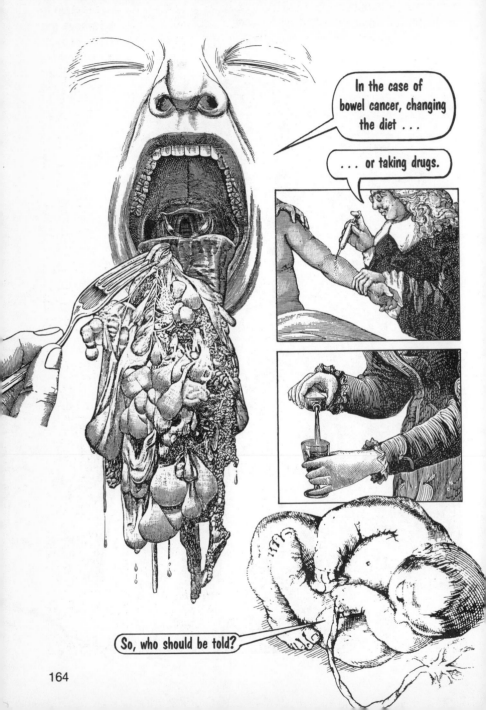

In the case of bowel cancer, changing the diet . . .

. . . or taking drugs.

So, who should be told?

This is difficult enough, but what about insurance? In the USA, people are being denied health cover because they have genes which may affect them later! To make a profit, the insurance companies prefer to take on good risks, leaving those with faulty genes to the tender mercies of the state.

And who do your genes belong to? Not, it seems, you. Now it is common to patent new pieces of human DNA as they are discovered. Some of these may be worth millions: perhaps they can be used as tests for inherited disease, or for very early cancer. But those who gave the gene — and perhaps had the disease — will not profit: the drug companies will.

Genetics has always turned out to be much more complicated than seemed reasonable to imagine.

And it is worth remembering that genetics had done a lot less to improve health than many people think. Twenty years ago, medical students learned almost no genetics — it seemed irrelevant to treating disease.

Today it is *still* irrelevant to treating most diseases, but students face scores of lectures on the subject. More is known about the genes for the red blood pigment than about any other — but this has been of almost no help in treating sickle cell anaemia.

This is what we need: more anatomy.

If nothing else, the history of genetics teaches humility. First — in spite of Galton — life is not simple.

Second, no-one is perfect. Almost everyone carries a potentially damaging gene, and most people die because of their own inborn imperfections.

Finally, genetics unites the human race — with itself, and with the rest of the living world. Human genetics, at its beginning, was a corrupt science which was widely abused. Now in its adolescence, it is ridding itself of its early problems; and will, perhaps, soon become a routine part of medicine.

But — never forget the past!

People mentioned in the text

Darwin, Charles Robert. *1809-1882.*
Fleeming Jenkin, Henry. *1833-1885.*
Galton, Sir Francis. *1822-1991.*
Mendel, Gregor Johann. *1822-1884.*
Joseph Merrick, the "Elephant Man". *1860-1890.*
Frederick II of Prussia, Frederick the Great.
 1712-1786.
Stopes, Marie Charlotte Carmichael. *1880-1958.*
Shaw, George Bernard. *1856-1950.*
Churchill, Sir Winston Leonard Spencer.
 1874-1965.
Hitler, Adolf. *1889-1945.*
de Vries, Hugo Marie. *1848-1935.*
Garrod, Archibald E. *1857-1936.*
Morgan, Thomas Hunt. *1866-1945.*
Watson, James Dewey. *1928-*
Crick, Francis Harry Compton. *1916-*
Sturtevant, A.H. *1891-*
Stalin, Joseph. *1879-1953.*
Lysenko, Trofim Denisovich. *1898-1976.*
Vavilov, Nikolai Ivanovich. *1887-1943.*
Muller, Hermann Joseph. *1890-*
Auerbach, Charlotte.
Miescher, Johann Friedrich. *1844-1895.*
Nägeli, C. *1817-1891.*
Avery, Oswald Theodore. *1877-1955.*
McLeod, C.M. *1909-*
McCarty, Maclyn. *1911-*
Franklin, Rosalind. *1920-1958.*
Pauling, Linus Carl. *1901-*
Meselson, Matthew Stanley. *1930-*
Stahl, Ernest Henry. *1929-*
Jacob, Francois. *1920-*
Monod, Jacques Lucien. *1910-1976.*
Sager, Ruth.
McClintock, Barbara. *1902-*
Lenin, Vladimir Ilyich. *1870-1924.*

Footnote

Genetics began by being ignored. Now it has the opposite problem. Mendel was dismissed because his work seemed unimportant, but today genes are everywhere and the public is fascinated by their promises and disturbed by their threats. Scientists have been quick to emphasize both. Not for nothing has it been said that the four letters of the genetic code have become H, Y, P and E.

The last decade's advances have been amazing. We have the complete sequence of the DNA letters of the 60,000 or so working genes needed to make a human being, and will soon have that of all the so-called "junk" DNA sequence (which may reveal that it does more than its name implies). 10,000 different diseases have an inherited component, and – in principle at least – we know the genes involved.

That raises both hopes and fears. For diseases controlled by single genes, such as sickle-cell anaemia or cystic fibrosis, it has become easier to identify both carriers and foetuses at risk. Because any gene can be damaged in many ways – for example, there are more than 1,000 known mutations for cystic fibrosis – the tests are not straightforward, and often the best that will be possible is to tell people that they *are* carriers, rather than to reassure them that they are not. The decisions as to whether to become pregnant or to continue with a pregnancy will, however, become somewhat easier as the tests become less ambiguous.

Tests are commercially available for genes predisposing to cystic fibrosis and breast cancer; and the development of DNA "chips" that can screen many genes at once means that more will soon be on sale. Medicine will have to deal more and more with those who have – rightly or wrongly – diagnosed themselves as at risk.

Most people, we now realize, die of a genetic disease, or at least of a disease with a genetic component. For some, it will become possible to tell them of their plight – but why should we want to do so? Sometimes, the information is helpful. Those who inherit a disposition towards certain forms of colon cancer, for example, can be helped by surgery long before the disease appears. For other illnesses, people at high risk can be warned to avoid an environment dangerous to them. Smoking is dangerous, but a few smokers get away with it. However, anyone who carries a changed form of an enzyme involved in clearing mucus from the lungs will certainly drown in their own spit if they smoke – and that might be enough to persuade them not to. However, knowledge can be dangerous, particularly when health insurance gets involved.

The most successful kind of medicine has always been prevention rather than cure. Genetics is no different, and the hope of replacing damaged DNA by gene therapy is still around the corner, where it has been for the past ten years. Genetic surgery – the ability to snip out pieces of DNA and move them to new places – has done remarkable things, but so far has done little to cure disease.

It might, though, help prevent the world's population from starving, at least according to enthusiasts for genetically modified (GM) foods. They may be right. It has proved remarkably easy to move plant genes around. Already there are crops that have been altered to make them resistant to parasites, or to artificial weedkillers (which means that the fields can be sprayed, leaving the crop unharmed). Commercial optimism has, in Europe if not the United States, been matched by public concerns about health risks. Why people are worried by the remote risk that GM foods *might* be dangerous to eat when they are happy to eat cheeseburgers that definitely *are*, mystifies scientists, but science is less important than what consumers are willing to accept. Unless attitudes change, the hope of putting genes for, say, essential nutrients into Third World crops will probably not be fulfilled.

If interfering with plants alarms society, to do the same with animals outrages a vocal part of it. We still know rather little about how a fertilized egg turns into an adult, with hundreds of different kinds of tissue, each bearing exactly the same genetic message but with jobs as different as brain cells and bone. Although it has long been possible to grow adult plants and even frogs from single cells, the notion that it might be possible to do so with mammals seemed a fantasy – until the birth of Dolly the sheep in 1997. Then, with the simple trick of inserting the nucleus from an adult cell into an emptied egg and allowing it to develop inside a foster-mother, a sheep was made without sex: it was cloned.

Cloned sheep or cows might be important in farming, and might be used to make multiple copies of animals with inserted human genes for proteins such as growth hormone (which are already used in "pharming", the production of valuable drugs in milk). The publicity that followed Dolly led to immediate condemnation of the idea of human cloning, often without much thought as to quite why it should be so horrific. After all, we are used to identical twins (who are clones of each other), so why should an artificial version cause such horror? In the end, again, public opinion moulds what science can do, and the prospect of cloning a human being seems remote.

And why might anyone want to do it? Claims of an army of identical Saddam Husseins verge on the silly, and others of replicating a loved child who died young also seem unlikely. However, the technique has great promise in medicine. Cells of the very early embryo (stem cells, as they are called) have the potential to divide into a variety of tissues, and can be grown – cloned – in the laboratory, or even manipulated with foreign genes. Perhaps they could make new skin or blood cells, or, in time, even whole organs. Because this involves the use of very early embryos, made perhaps by artificial fertilization in the laboratory and not needed for implantation into

174

a mother, it has become mixed up with the abortion debate. In the United States, the "Pro-Life" lobby has succeeded in denying funds from government sources for such work.

Genetics is always mixed up with politics. It has been used both to blame and to excuse human behaviour. The claim (in the end not confirmed) of a "gay gene" led to two distinct responses among the homosexual community. Some feared that the gene would be used to stigmatize them, but most welcomed the idea that their behaviour might be coded into DNA, as it meant that they could not be accused of corrupting those not already "at risk". Such opposing views apply just as much to the supposed genes that predispose to crime – are they evidence that the criminal cannot be reformed and must be locked away for ever, or should they be used in mitigation to argue that he was not acting according to his own free will?

Science has no answer to such questions, and in the end the most surprising result of the new genetics may be how little it tells us about ourselves.

Further Reading

There are many excellent textbooks on genetics. One of the best is *An Introduction to Genetic Analysis* by Anthony Griffiths and others (Freeman, 1998). Matt Ridley's book *Genome: the Autobiography of a Species in 23 Chapters* (Fourth Estate, 1999) is a first-rate general introduction to the subject. Finally, for one man's view of human genetics and evolution, there are Steve Jones' own books *The Language of the Genes* (HarperCollins, 1993) and *In the Blood* (HarperCollins, 1997).

About the Author

Steve Jones is Professor of Genetics at University College London. He obtained his first degree and PhD at the University of Edinburgh, and has held positions in various universities in Britain, the United States, Africa and Australia. He gave the 1991 BBC Radio Reith Lectures on "The Language of the Genes", and his book of the same title was published in 1993.

About the Illustrator

Borin Van Loon has illustrated seven other titles in this series: *Introducing Mathematics, Sociology, Cultural Studies, Media Studies, Buddha, Eastern Philosophy* and *Darwin and Evolution*. He is a surrealist artist whose work ranges from oil paintings to a cut-out book on DNA.

Index

abortion 151–3
albinism 26
amino acids 48, 64–5
animals, genetic
  engineering 143
antigens 149
atavism 25
Auerbach, Charlotte 45

bacterial mating 67–71,
  79–82
blood cells, red 87–9
brain genes 124

cancer 146–7, 162–3
cells 74–5, 87–9
chromosomes 34–44, 100
  bacterial genes 67–71
  defined 77
  see also DNA; YACs
Churchill, Winston 28
cloning 81, 174–5
codes 77
conception 75
Crick, Francis 54–63
criminal tendency 156
cystic fibrosis 117–21, 153
  screening for 132–5
  treating 137–9, 144

Darwin, Charles 8–9
disease, predicting see
  screening
disease, inherited 114–23,
  128–53, 158, 173–4
DNA 51ff
  artificial 64
  codes 53–65, 77
  in conception 75
  creating 79–85
  heredity 76–7
  length 73
  mapping 78–113
  migratory 101–5
dominant genes 19, 33
double helix 56–9

egg, human 75
electrophoresis 84
enzymes 31, 77
  see also polymerases;
  restriction enzymes
Eugenics Society 27
eye colour 33–8

female sex (prejudice) 157
Franklin, Rosalind 56
fruit flies, study of 32–9,
  104
  mutations 44–6

Galton, Francis 11–13, 156
Galton Laboratory 27
Garrod, Archibald 31
genes
  defined 5
  dominant/recessive 19
  linkage map 40–1, 78
  replacing faulty 144–5
  transferring 166
genetic
  diseases, curing 11
  engineering 139–65
  map 122, 125–7, 136
genetically modified foods
  174
germ-line therapy 150–1,
  154

haemoglobin 87–8
heredity 76
Hitler, Adolf 28
HIV virus 111
human
  egg 75
  inheritance 23
Huntington's Disease 160

inheritance, human 23
insurance, health 165, 173
intelligence 155

Jenkin, Fleeming 9–10

kidney failure 161

Laboratory for National
  Eugenics 13, 27
linkage map 40–1, 78
Lysenko, T.D. 42–3

mapping
  DNA 78–113
  genes 122
McClintock, Barbara 101
Mendel, Gregor 14–23
mice 148
Miescher, J.F. 46–7
Morgan, Thomas Hunt 32–8
mRNA 63, 77
mutations 30, 44–6, 60,
  66, 101–5

nucleic acids 47–51

operons 70

Pauling, Linus 56
PCR 83
peas 15–21
pedigrees 24, 26
phages 52, 61
plasmids 81
pneumonia 49
politics 41, 174–5
polymerases 60, 83
prions 108–9
proteins 47–8, 77

recessive genes 19, 25,
  26, 33
restriction enzymes 80
retroviruses 110–11
RNA 51, 63
  see also mRNA, tRNA

Sager, Ruth 73
screening 133–6, 158–9
sex cells 75, 76
Shaw, G.B. 28
sickle-cell anaemia 158
Stalin, Joseph 42
starvation, easing 174
stem cells 174–5
Stopes, Marie 27
Sturtevant, A.H. 39–40

Tay-Sachs Disease 115,
  131
transcription 63
translation 63
transforming principle 50
tRNA 77

variation in genetics 29
viruses 52, 80–1, 106–9
  genetic engineering 142
  see also HIV virus;
  retroviruses
Vries, Hugo de 30

Watson, James 54–63
wobble theory 65

X-rays 45–6
  and DNA 56

YACs (yeast artificial
  chromosomes) 82